T0253956

Lecture Notes in Computer Science　　10044

Commenced Publication in 1973
Founding and Former Series Editors:
Gerhard Goos, Juris Hartmanis, and Jan van Leeuwen

More information about this series at http://www.springer.com/series/7409

Tilmann Rabl · Raghunath Nambiar
Chaitanya Baru · Milind Bhandarkar
Meikel Poess · Saumyadipta Pyne (Eds.)

Big Data Benchmarking

6th International Workshop, WBDB 2015
Toronto, ON, Canada, June 16–17, 2015 and
7th International Workshop, WBDB 2015
New Delhi, India, December 14–15, 2015
Revised Selected Papers

 Springer

Editors
Tilmann Rabl
Technical University of Berlin
Berlin
Germany

Raghunath Nambiar
Cisco Systems, Inc.
San Jose, CA
USA

Chaitanya Baru
University of California at San Diego
La Jolla, CA
USA

Milind Bhandarkar
Ampool, Inc.
Santa Clara, CA
USA

Meikel Poess
Oracle Corporation
Redwood Shores, CA
USA

Saumyadipta Pyne
Indian Institute of Public Health
Hyderabad
India

ISSN 0302-9743   ISSN 1611-3349 (electronic)
Lecture Notes in Computer Science
ISBN 978-3-319-49747-1   ISBN 978-3-319-49748-8 (eBook)
DOI 10.1007/978-3-319-49748-8

Library of Congress Control Number: 2016958990

LNCS Sublibrary: SL3 – Information Systems and Applications, incl. Internet/Web, and HCI

This Springer imprint is published by Springer Nature
The registered company is Springer International Publishing AG
The registered company address is: Gewerbestrasse 11, 6330 Cham, Switzerland

Preface

Big data processing has evolved from a buzz-word to a pervasive technology shift in only a few years. The plethora of systems developed or extended for big data processing makes big data benchmarking more important than ever.

Formed in 2012, the Big Data Benchmarking Community (BDBC) represented a major step in facilitating the development of benchmarks for objective comparisons of hardware and software systems dealing with emerging big data applications. Led by Chaitanya Baru, Tilmann Rabl, Milind Bhandarkar, Raghunath Nambiar, and Meikel Poess, the BDBC has successfully conducted seven international Workshops on Big Data Benchmarking (WBDB) in only 4 years bringing industry experts and researchers together to present and debate challenges, ideas, and methodologies to benchmark big data systems. In 2015, the BBDC joined the SPEC Research Group to further organize and structure the benchmarking efforts collected under the BBDC umbrella. The Big Data Working Group[1] meets regularly as a platform for discussion and development of big data benchmarks.

While being a platform of discussion, the BBDC and SPEC Research Group also demonstrated practical impact in the field of big data benchmarking. In 2015, Big-Bench, one of the benchmark efforts started at WBDB2012.us, was adopted by the Transaction Processing Performance Council to become the first industry standard end-to-end benchmark for big data analysis. The benchmark was officially released under the name TPCx-BB in January 2016[2].

This book contains the joint proceedings of the 6th and 7th Workshop on Big Data Benchmarking. The 6th WBDB was held in Toronto, Canada, during June 16–17, 2015, hosted by the University of Toronto. The 7th WBDB was held in New Delhi, India, during December 14–15, 2015, at the India Habitat Centre. Both workshops were well attended by industry and academia alike and featured SPEC Research Big Data Working Group meetings. Keynote speakers in WBDB2015.ca were Tamer Özsu (University of Waterloo) and Anil Goel (SAP). Tamer Özsu presented benchmarks for graph management and Anil Goel gave an overview of the SAP big data infrastructure. In WBDB2015.in Michael Franklin (UC Berkeley) presented emerging trends in big data software, Geoffrey Fox (Indiana University Bloomington) discussed the convergence of big data and high-performance computing (HPC), and Mukund Desphande (Persistent Systems) introduced the use of Web technologies in database architecture.

In this book, we have collected recent trends in big data and HPC convergence, new proposals for big data benchmarking, as well as tooling and performance results. In the first part of the book, an overview of challenges and opportunities of the convergence of HPC and big data is given. The second part presents two novel benchmarks, one based on the Indian Aadhaar database that contains biometric information of hundreds

---

[1] See https://research.spec.org/working-groups/big-data-working-group.

[2] See http://www.tpc.org/tpcx-bb/default.asp.

of millions of people and one for array database systems. The third part contains experiences from a benchmarking framework for Hadoop setups. In the final part, several performance results from extensive experiments on Cassandra, Spark, and Hadoop are presented.

The seven papers in this book were selected out of a total of 39 presentations in WBDB2015.ca and WBDB2015.in. All papers were reviewed in two rounds. We thank the sponsors, members of the Program Committee, authors, and participants for their contributions to these workshops. The hard work and close cooperation of a number of people have been critical to the success of the WBDB workshop series.

August 2016   Tilmann Rabl

WBDB 2015 Organization

General Chairs

Tilmann Rabl   Technische Universität Berlin, Germany
Chaitanya Baru   San Diego Supercomputer Center, USA

Program Committee Chairs

WBDB2015.ca

Tilmann Rabl   Technische Universität Berlin, Germany

WBDB2015.in

Saumyadipta Pyne   Indian Institute of Public Health, Hyderabad, India

Program Committee

Nelson Amaral   University of Alberta, Canada
Amitabha Bagchi   IIT Delhi, India
Milind Bhandarkar   Ampool, USA
Tobias Bürger   Payback, Germany
Ann Cavoukian   Ryerson University, Canada
Fei Chiang   McMaster University, Canada
Pedro Furtado   University of Coimbra, Portugal
Pulak Ghosh   IIM Bangalore, India
Boris Glavic   IIT Chicago, USA
Parke Godfrey   York University, Canada
Anil Goel   SAP, Canada
Bhaskar Gowda   Intel, USA
Chetan Gupta   Hitachi, USA
Rajendra Joshi   C-DAC, Pune, India
Patrick Hung   University of Ontario Institute of Technology, Canada
Arnab Laha   IIM Ahmedabad, India
Asterios Katsifodimos   TU Berlin, Germany
Jian Li   Huawei, Canada
Serge Mankovskii   CA, USA
Tridib Mukherjee   Xerox Research Center India
Raghunath Nambiar   Cisco, USA
Glenn Paulley   Conestoga College, Canada
Scott Pearson   Cray, USA
Meikel Poess   Oracle, USA

Nicolas Poggi	Barcelona Supercomputer Center, Spain
Saumyadipta Pyne	CRRao AIMSCS, Hyderabad, India
Francois Raab	InfoSizing, USA
V.C.V. Rao	C-DAC, Pune, India
Ken Salem	University of Waterloo, Canada
Berni Schiefer	IBM, USA
Yogesh Simmhan	IISc, USA
Raghavendra Singh	IBM, India
Dinkar Sitaram	PES University, Bangalore, India
Chiranjib Sur	Shell, India
Ali Tizghadam	Telus, Canada
Jonas Traub	Technische Universität Berlin, Germany
Matthias Uflacker	Hasso Plattner Institut, Germany
Herna Viktor	University of Ottawa, Canada
Anil Vullikanti	Virginia Bioinformatics Institute, USA
Jianfeng Zhan	Chinese Academy of Sciences, China
Roberto V. Zicari	Frankfurt Big Data Lab, Goethe University Frankfurt, Germany

WBDB 2015 Sponsors

WBDB2015.ca Sponsors

Platinum Sponsor	Intel
Gold Sponsor	IBM
Silver Sponsor	Cray

WBDB2015.in Sponsors

Gold Sponsors	Intel, NSF
Silver Sponsors	Ampool, Government of India
Bronze Sponsors	Persistent
Dinner Sponsor	Nvidia

Contents

Future Challenges

Big Data, Simulations and HPC Convergence

Geoffrey Fox[1]([✉]), Judy Qiu[1], Shantenu Jha[2],
Saliya Ekanayake[1], and Supun Kamburugamuve[1]

[1] School of Informatics and Computing, Indiana University,
Bloomington, IN 47408, USA
gcf@indiana.edu
[2] ECE, Rutgers University, Piscataway, NJ 08854, USA

Abstract. Two major trends in computing systems are the growth in high performance computing (HPC) with in particular an international exascale initiative, and big data with an accompanying cloud infrastructure of dramatic and increasing size and sophistication. In this paper, we study an approach to convergence for software and applications/algorithms and show what hardware architectures it suggests. We start by dividing applications into data plus model components and classifying each component (whether from Big Data or Big Compute) in the same way. This leads to 64 properties divided into 4 views, which are Problem Architecture (Macro pattern); Execution Features (Micro patterns); Data Source and Style; and finally the Processing (runtime) View. We discuss convergence software built around HPC-ABDS (High Performance Computing enhanced Apache Big Data Stack) and show how one can merge Big Data and HPC (Big Simulation) concepts into a single stack and discuss appropriate hardware.

Keywords: Big Data · HPC · Simulations

1 Introduction

Two major trends in computing systems are the growth in high performance computing (HPC) with an international exascale initiative, and the big data phenomenon with an accompanying cloud infrastructure of well publicized dramatic and increasing size and sophistication. There has been substantial discussion of the convergence of big data analytics, simulations and HPC [1, 11–13, 29, 30] highlighted by the Presidential National Strategic Computing Initiative [5]. In studying and linking these trends and their convergence, one needs to consider multiple aspects: hardware, software, applications/algorithms and even broader issues like business model and education. Here we focus on software and applications/algorithms and make comments on the other aspects. We discuss applications/algorithms in Sect. 2, software in Sect. 3 and link them and other aspects in Sect. 4.

© Springer International Publishing AG 2016
T. Rabl et al. (Eds.): WBDB 2015, LNCS 10044, pp. 3–17, 2016.
DOI: 10.1007/978-3-319-49748-8_1

2   Applications and Algorithms

We extend the analysis given by us [18,21], which used ideas in earlier parallel computing studies [8,9,31] to build a set of Big Data application characteristics with 50 features – called facets – divided into 4 views. As it incorporated the approach of the Berkeley dwarfs [8] and included features from the NRC Massive Data Analysis Reports Computational Giants [27], we termed these characteristics as Ogres. Here we generalize approach to integrate Big Data and Simulation applications into a single classification that we call convergence diamonds with a total of 64 facets split between the same 4 views. The four views are Problem Architecture (Macro pattern abbreviated PA); Execution Features (Micro patterns abbreviated EF); Data Source and Style (abbreviated DV); and finally the Processing (runtime abbreviated Pr) View.

The central idea is that any problem – whether Big Data or Simulation, and whether HPC or cloud-based, can be broken up into Data plus Model. The DDDAS approach is an example where this idea is explicit [3]. In a Big Data problem, the Data is large and needs to be collected, stored, managed and accessed. Then one uses Data Analytics to compare some Model with this data. The Model could be small such as coordinates of a few clusters or large as in a deep learning network; almost by definition the Data is large!

On the other hand for simulations the model is nearly always big – as in values of fields on a large space-time mesh. The Data could be small and is essentially zero for Quantum Chromodynamics simulations and corresponds to the typically small boundary conditions for many simulations; however climate and weather simulations can absorb large amounts of assimilated data. Remember Big Data has a model, so there are model diamonds for big data. The diamonds and their facets are given in a table put in the appendix. They are summarized above in Fig. 1.

Comparing Big Data and simulations is not so clear; however comparing the model in simulations and the model in Big Data is straightforward while the data in both cases can be treated similarly. This simple idea lies at heart of our approach to Big Data - Simulation convergence. In the convergence diamonds given in Table presented in Appendix, one divides the facets into three types

1. Facet n (without D or M) refers to a facet of system including both data and model – 16 in total.
2. Facet nD is a Data only facet – 16 in Total
3. Facet nM is a Model only facet – 32 in total

The increase in total facets and large number of model facets corresponds mainly to adding Simulation facets to the Processing View of the Diamonds. Note we have included characteristics (facets) present in the Berkeley Dwarfs and NAS Parallel Benchmarks as well the NRC Massive Data Analysis Computational Giants. For some facets there are separate data and model facets. A good example in "Convergence Diamond Micropatterns or Execution Features" is that EF-4D is Data Volume and EF-4M Model size.

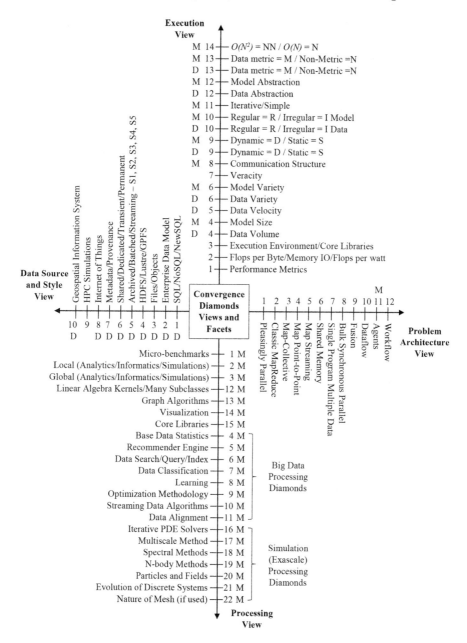

Fig. 1. Summary of the 64 facets in the convergence diamonds

The views Problem Architecture; Execution Features; Data Source and Style; and Processing (runtime) are respectively mainly System Facets, a mix of system, model and data facets; mainly data facets with the final view entirely model facets. The facets tell us how to compare diamonds (instances of big data and simulation applications) and see which system architectures are needed to support each diamond and which architectures across multiple diamonds including those from both simulation and big data areas.

In several papers [17,33,34] we have looked at the model in big data problems and studied the model performance on both cloud and HPC systems. We have shown similarities and differences between models in simulation and big data area. In particular the latter often need HPC hardware and software enhancements to get good performances. There are special features of each class; for example simulations often have local connections between model points corresponding either to the discretization of a differential operator or a short range force. Big data sometimes involve fully connected sets of points and these formulations have similarities to long range force problems in simulation. In both regimes we often see linear algebra kernels but the sparseness structure is rather different. Graph data structures are present in both cases but that in simulations tends to have more structure. The linkage between people in Facebook social network is less structured than the linkage between molecules in a complex biochemistry simulation. However both are graphs with some long range but many short range interactions. Simulations nearly always involve a mix of point to point messaging and collective operations like broadcast, gather, scatter and reduction. Big data problems sometimes are dominated by collectives as opposed to point to point messaging and this motivates the map collective problem architecture facet PA-3 above. In simulations and big data, one sees a similar BSP (loosely synchronous PA-8), SPMD (PA-7) Iterative (EF-11M) and this motivates the Spark [32], Flink [7], Twister [15,16] approach. Note that pleasingly parallel (PA-1) local (Pr-2M) structure is often seen in both simulations and big data.

In [33,34] we introduce Harp as a plug-in to Hadoop with scientific data abstractions, support of iterations and high quality communication primitives. This runs with good performance on several important data analytics including Latent Dirichlet Allocation LDA, clustering and dimension reduction. Note LDA has a non trivial structure sparse structure coming from an underlying bag of words model for documents. In [17], we look at performance in great detail showing excellent data analytics speed up on an Infiniband connected HPC cluster using MPI. Deep Learning [14,24] has clearly shown importance of HPC and uses many ideas originally developed for simulations.

Above we discuss models in the big data and simulation regimes; what about the data? Here we see the issue as perhaps less clear but convergence does not seem difficult technically. Given models can be executed on HPC systems when needed, it appears reasonable to use a different architecture for the data with the big data approach of hosting data on clouds quite attractive. HPC has tended not to use big data management tools but rather to host data on shared file systems like Lustre. We expect this to change with object stores and HDFS approaches gaining popularity in the HPC community. It is not clear if HDFS will run on HPC systems or instead on co-located clouds supporting the rich object, SQL, NoSQL and NewSQL paradigms. This co-location strategy can also work for streaming data with in the traditional Apache Storm-Kafka map streaming model (PA-5) buffering data with Kafka on a cloud and feeding that data to Apache Storm that may need HPC hardware for complex analytics (running on bolts in Storm). In this regard we have introduced HPC enhancements to Storm [26].

We believe there is an immediate need to investigate the overlap of application characteristics and classification from high-end computing and big data ends of the spectrum. Here we have shown how initial work [21] to classify big data applications can be extended to include traditional high-performance applications. Can traditional classifications for high-performance applications [8] be extended in the opposite direction to incorporate big data applications? And if so, is the end result similar, overlapping or very distinct to the preliminary classification proposed here? Such understanding is critical in order to eventually have a common set of benchmark applications and suites [10] that will guide the development of future systems that must have a design point that provides balanced performance.

Note applications are instances of Convergence Diamonds. Each instance will exhibit some but not all of the facets of Fig. 1. We can give an example of the NAS Parallel Benchmark [4] LU (Lower-Upper symmetric Gauss Seidel) using MPI. This would be a diamond with facets PA-4, 7, 8; Pr-3M,16M with its size specified in EF-4M. PA-4 would be replaced by PA-2 if one used (unwisely) MapReduce for this problem. Further if you read initial data from MongoDB, the data facet DV-1D would be added. Many other examples are given in Sect. 3 of [18]. For example non-vector clustering in Table 1 of this section is a nice data analytics example. It exhibits Problem Architecture view PA-3, PA-7, and PA-8; Execution Features EF-9D (Static), EF-10D (Regular), EF-11M (iterative), EF-12M (bag of items), EF-13D (Non-metric), EF-13M(Non metric), and EF-14M($O(N2)$ algorithm); Processing view Pr-3M, Pr-9M (Machine learning and Expectation maximization), and Pr-12M (Full matrix, Conjugate Gradient).

3   HPC-ABDS Convergence Software

In previous papers [20, 25, 28], we introduced the software stack HPC-ABDS (High Performance Computing enhanced Apache Big Data Stack) shown online [4] and in Figs. 2 and 3. These were combined with the big data application analysis [6, 19, 21] in terms of Ogres that motivated the extended convergence diamonds in Sect. 2. We also use Ogres and HPC-ABDS to suggest a systematic approach to benchmarking [18, 22]. In [23] we described the software model of Fig. 2 while further details of the stack can be found in an online course [2] that includes a section with about one slide (and associated lecture video) for each entry in Fig. 2.

Figure 2 collects together much existing relevant systems software coming from either HPC or commodity sources. The software is broken up into layers so software systems are grouped by functionality. The layers where there is especial opportunity to integrate HPC and ABDS are colored green in Fig. 2. This is termed HPC-ABDS (High Performance Computing enhanced Apache Big Data Stack) as many critical core components of the commodity stack (such as

Kaleidoscope of (Apache) Big Data Stack (ABDS) and HPC Technologies	
Cross-Cutting Functions	17) **Workflow-Orchestration:** ODE, ActiveBPEL, Airavata, Pegasus, Kepler, Swift, Taverna, Triana, Trident, BioKepler, Galaxy, IPython, Dryad, Naiad, Oozie, Tez, Google FlumeJava, Crunch, Cascading, Scalding, e-Science Central, Azure Data Factory, Google Cloud Dataflow, NiFi (NSA), Jitterbit, Talend, Pentaho, Apatar, Docker Compose, KeystoneML
1) **Message and Data Protocols:** Avro, Thrift, Protobuf	16) **Application and Analytics:** Mahout , MLlib , MLbase, DataFu, R, pbdR, Bioconductor, ImageJ, OpenCV, Scalapack, PetSc, PLASMA MAGMA, Azure Machine Learning, Google Prediction API & Translation API, mlpy, scikit-learn, PyBrain, CompLearn, DAAL(Intel), Caffe, Torch, Theano, DL4j, H2O, IBM Watson, Oracle PGX, GraphLab, GraphX, IBM System G, GraphBuilder(Intel), TinkerPop, Parasol, Dream:Lab, Google Fusion Tables, CINET, NWB, Elasticsearch, Kibana, Logstash, Graylog, Splunk, Tableau, D3.js, three.js, Potree, DC.js, TensorFlow, CNTK
2) **Distributed Coordination:** Google Chubby, Zookeeper, Giraffe, JGroups	15B) **Application Hosting Frameworks:** Google App Engine, AppScale, Red Hat OpenShift, Heroku, Aerobatic, AWS Elastic Beanstalk, Azure, Cloud Foundry, Pivotal, IBM BlueMix, Ninefold, Jelastic, Stackato, appfog, CloudBees, Engine Yard, CloudControl, dotCloud, Dokku, OSGi, HUBzero, OODT, Agave, Atmosphere
	15A) **High level Programming:** Kite, Hive, HCatalog, Tajo, Shark, Phoenix, Impala, MRQL, SAP HANA, HadoopDB, PolyBase, Pivotal HD/Hawq, Presto, Google Dremel, Google BigQuery, Amazon Redshift, Drill, Kyoto Cabinet, Pig, Sawzall, Google Cloud DataFlow, Summingbird
3) **Security & Privacy:** InCommon, Eduroam, OpenStack, Keystone, LDAP, Sentry, Sqrrl, OpenID, SAML OAuth	14B) **Streams:** Storm, S4, Samza, Granules, Neptune, Google MillWheel, Amazon Kinesis, LinkedIn, Twitter Heron, Databus, Facebook Puma/Ptail/Scribe/ODS, Azure Stream Analytics, Floe, Spark Streaming, Flink Streaming, DataTurbine
	14A) **Basic Programming model and runtime, SPMD, MapReduce:** Hadoop, Spark, Twister, MR-MPI, Stratosphere (Apache Flink), Reef, Disco, Hama, Giraph, Pregel, Pegasus, Ligra, GraphChi, Galois, Medusa-GPU, MapGraph, Totem
	13) **Inter process communication Collectives, point-to-point, publish-subscribe:** MPI, HPX-5, Argo BEAST HPX-5 BEAST PULSAR, Harp, Netty, ZeroMQ, ActiveMQ, RabbitMQ, NaradaBrokering, QPid, Kafka, Kestrel, JMS, AMQP, Stomp, MQTT, Marionette Collective, **Public Cloud:** Amazon SNS, Lambda, Google Pub Sub, Azure Queues, Event Hubs
4) **Monitoring:** Ambari, Ganglia, Nagios, Inca	12) **In-memory databases/caches:** Gora (general object from NoSQL), Memcached, Redis, LMDB (key value), Hazelcast, Ehcache, Infinispan, VoltDB, H-Store
	12) **Object-relational mapping:** Hibernate, OpenJPA, EclipseLink, DataNucleus, ODBC/JDBC
	12) **Extraction Tools:** UIMA, Tika
21 layers **Over 350** **Software** **Packages** **January 29** **2016**	11C) **SQL(NewSQL):** Oracle, DB2, SQL Server, SQLite, MySQL, PostgreSQL, CUBRID, Galera Cluster, SciDB, Rasdaman, Apache Derby, Pivotal Greenplum, Google Cloud SQL, Azure SQL, Amazon RDS, Google F1, IBM dashDB, N1QL, BlinkDB, Spark SQL
	11B) **NoSQL:** Lucene, Solr, Solandra, Voldemort, Riak, ZHT, Berkeley DB, Kyoto/Tokyo Cabinet, Tycoon, Tyrant, MongoDB, Espresso, CouchDB, Couchbase, IBM Cloudant, Pivotal Gemfire, HBase, Google Bigtable, LevelDB, Megastore and Spanner, Accumulo, Cassandra, RYA, Sqrrl, Neo4J, graphdb, Yarcdata, AllegroGraph, Blazegraph, Facebook Tao, Titan:db, Jena, Sesame **Public Cloud:** Azure Table, Amazon Dynamo, Google DataStore
	11A) **File management:** iRODS, NetCDF, CDF, HDF, OPeNDAP, FITS, RCFile, ORC, Parquet
	10) **Data Transport:** BitTorrent, HTTP, FTP, SSH, Globus Online (GridFTP), Flume, Sqoop, Pivotal GPLOAD/GPFDIST
	9) **Cluster Resource Management**: Mesos, Yarn, Helix, Llama, Google Omega, Facebook Corona, Celery, HTCondor, SGE, OpenPBS, Moab, Slurm, Torque, Globus Tools, Pilot Jobs
	8) **File systems:** HDFS, Swift, Haystack, f4, Cinder, Ceph, FUSE, Gluster, Lustre, GPFS, GFFS **Public Cloud:** Amazon S3, Azure Blob, Google Cloud Storage
	7) **Interoperability:** Libvirt, Libcloud, JClouds, TOSCA, OCCI, CDMI, Whirr, Saga, Genesis
	6) **DevOps:** Docker (Machine, Swarm), Puppet, Chef, Ansible, SaltStack, Boto, Cobbler, Xcat, Razor, CloudMesh, Juju, Foreman, OpenStack Heat, Sahara, Rocks, Cisco Intelligent Automation for Cloud, Ubuntu MaaS, Facebook Tupperware, AWS OpsWorks, OpenStack Ironic, Google Kubernetes, Buildstep, Gitreceive, OpenTOSCA, Winery, CloudML, Blueprints, Terraform, DevOpSlang, Any2Api
	5) **IaaS Management from HPC to hypervisors:** Xen, KVM, QEMU, Hyper-V, VirtualBox, OpenVZ, LXC, Linux-Vserver, OpenStack, OpenNebula, Eucalyptus, Nimbus, CloudStack, CoreOS, rkt, VMware ESXi, vSphere and vCloud, Amazon, Azure, Google and other public Clouds **Networking:** Google Cloud DNS, Amazon Route 53

Fig. 2. Big Data and HPC Software subsystems arranged in 21 layers. Green layers have a significant HPC integration (Color figure online)

Spark and Hbase) come from open source projects while HPC is needed to bring performance and other parallel computing capabilities [23]. Note that Apache is the largest but not only source of open source software; we believe that the Apache Foundation is a critical leader in the Big Data open source software movement and use it to designate the full big data software ecosystem. The figure also includes proprietary systems as they illustrate key capabilities and often

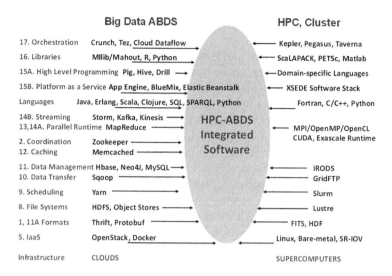

Fig. 3. Comparison of Big Data and HPC simulation software stacks

motivate open source equivalents. We built this picture for big data problems but it also applies to big simulation with caveat that we need to add more high level software at the library level and more high level tools like Global Arrays. This will become clearer in the next section when we discuss Fig. 2 in more detail.

The essential idea of our Big Data HPC convergence for software is to make use of ABDS software where possible as it offers richness in functionality, a compelling open-source community sustainability model and typically attractive user interfaces. ABDS has a good reputation for scale but often does not give good performance. We suggest augmenting ABDS with HPC ideas especially in the green layers of Fig. 2. We have illustrated this with Hadoop [33,34], Storm [26] and the basic Java environment [17]. We suggest using the resultant HPC-ABDS for both big data and big simulation applications. In the language of Fig. 2, we use the stack on left enhanced by the high performance ideas and libraries of the classic HPC stack on the right. As one example we recommend using enhanced MapReduce (Hadoop, Spark, Flink) for parallel programming for simulations and big data where it's the model (data analytics) that has similar requirements to simulations. We have shown how to integrate HPC technologies into MapReduce to get performance expected in HPC [34] and that on the other hand if the user interface is not critical, one can use a simulation technology (MPI) to drive excellent data analytics performance [17]. A byproduct of these studies is that classic HPC clusters make excellent data analytics engine. One can use the convergence diamonds to quantify this result. These define properties of applications between both data and simulations and allow one to specify hardware and software requirements uniformly over these two classes of applications.

4   Convergence Systems

Figure 3 contrasts modern ABDS and HPC stacks illustrating most of the 21 layers and labelling on left with layer number used in Fig. 2. The omitted layers in Fig. 2 are Interoperability, DevOps, Monitoring and Security (layers 7, 6, 4, 3) which are all important and clearly applicable to both HPC and ABDS. We also add in Fig. 3, an extra layer corresponding to programming language, which feature is not discussed in Fig. 2. Our suggested approach is to build around the stacks of Fig. 2, taking the best approach at each layer which may require merging ideas from ABDS and HPC. This converged stack is still emerging but we have described some features in the previous section. Then this stack would do both big data and big simulation as well the data aspects (store, manage, access) of the data in the "data plus model" framework. Although the stack architecture is uniform it will have different emphases in hardware and software that will be optimized using the convergence diamond facets. In particular the data management will usually have a different optimization from the model computation.

Thus we propose a canonical "dual" system architecture sketched in Fig. 4 with data management on the left side and model computation on the right. As drawn the systems are the same size but this of course need not be true. Further we depict data rich nodes on left to support HDFS but that also might not be correct – maybe both systems are disk rich or maybe we have a classic Lustre style system on the model side to mimic current HPC practice. Finally the systems may in fact be coincident with data management and model computation on the same nodes. The latter is perhaps the canonical big data approach but we see many big data cases where the model will require hardware optimized for performance and with for example high speed internal networking or GPU enhanced nodes. In this case the data may be more effectively handled by a separate cloud like cluster. This depends on properties recorded in the facets of the Convergence Diamonds for application suites. These ideas are built on substantial experimentation but still need significant testing as they have not be looked at systematically.

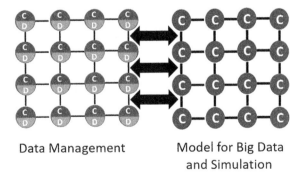

Data Management   Model for Big Data
                         and Simulation

Fig. 4. Dual convergence architecture

We suggested using the same software stack for both systems in the dual Convergence system. Now that means we pick and chose from HPC-ABDS on both machines but we needn't make same choice on both systems; obviously the data management system would stress software in layers 10 and 11 of Fig. 2 while the model computation would need libraries (layer 16) and programming plus communication (layers 13–15).

Acknowledgments. This work was partially supported by NSF CIF21 DIBBS 1443054, NSF OCI 1149432 CAREER. and AFOSR FA9550-13-1-0225 awards. We thank Dennis Gannon for comments on an early draft.

Appendix: Convergence Diamonds with 64 Facets

These are discussed in Sect. 2 and summarized in Fig. 1

Table 1. Convergence Diamonds and their Facets.

Facet and View		Comments
PA: Problem Architecture View of Diamonds (Meta or MacroPatterns)		
Nearly all are the system of Data and Model		
PA-1	Pleasingly Parallel	As in BLAST, Protein docking. Includes Local Analytics or Machine Learning ML or filtering pleasingly parallel, as in bio-imagery, radar images (pleasingly parallel but sophisticated local analytics)
PA-2	Classic MapReduce	Search, Index and Query and Classification algorithms like collaborative filtering
PA-3	Map-Collective	Iterative maps + communication dominated by "collective" operations as in reduction, broadcast, gather, scatter. Common datamining pattern but also seen in simulations
4	Map Point-to-Point	Iterative maps + communication dominated by many small point to point messages as in graph algorithms and simulations
PA-5	Map Streaming	Describes streaming, steering and assimilation problems
PA-6	Shared memory (as opposed to distributed parallel algorithm)	Corresponds to problem where shared memory implementations important. Tend to be dynamic and asynchronous
PA-7	SPMD	Single Program Multiple Data, common parallel programming feature
PA-8	Bulk Synchronous Processing (BSP)	Well-defined compute-communication phases
PA-9	Fusion	Full applications often involves fusion of multiple methods. Only present for composite Diamonds
PA-10	Dataflow	Important application features often occurring in composite Diamonds
PA-11M	Agents	Modelling technique used in areas like epidemiology (swarm approaches)
PA-12	Orchestration (workflow)	All applications often involve orchestration (workflow) of multiple components

Table 1. *Continued*

EF: Diamond Micropatterns or Execution Features		
EF-1	Performance Metrics	Result of Benchmark
EF-2	Flops/byte (Memory or I/O). Flops/watt (power)	I/O Not needed for "pure in memory" benchmark
EF-3	Execution Environment	Core libraries needed: matrix-matrix/vector algebra, conjugate gradient, reduction, broadcast; Cloud, HPC, threads, message passing etc. Could include details of machine used for benchmarking here
EF-4D	Data Volume	Property of a Diamond Instance. Benchmark measure
EF-4M	Model Size	
EF-5D	Data Velocity	Associated with streaming facet but value depends on particular problem. Not applicable to model
EF-6D	Data Variety	Most useful for composite Diamonds. Applies separately for model and data
EF-6M	Model Variety	
EF-7	Veracity	Most problems would not discuss but potentially important
EF-8M	Communication Structure	Interconnect requirements; Is communication BSP, Asynchronous, Pub-Sub, Collective, Point to Point? Distribution and Synch
EF-9D	D = Dynamic or S = Static Data	Clear qualitative properties. Importance familiar from parallel computing and important separately for data and model
EF-9M	D = Dynamic or S = Static Model	Clear qualitative properties. Importance familiar from parallel computing and important separately for data and model
EF-10D	R = Regular or I = Irregular Data	
EF-10M	R = Regular or I = Irregular Model	
EF-11M	Iterative or not?	Clear qualitative property of Model. Highlighted by Iterative MapReduce and always present in classic parallel computing
EF-12D	Data Abstraction	e.g. key-value, pixel, graph, vector, bags of words or items. Clear quantitative property although important data abstractions not agreed upon. All should be supported by Programming model and run time
EF-12M	Model Abstraction	e.g. mesh points, finite element, Convolutional Network.
EF-13D	Data in Metric Space or not?	Important property of data.
EF-13M	Model in Metric Space or not?	Often driven by data but model and data can be different here
EF-14M	$O(N^2)$ or $O(N)$ Complexity?	Property of Model algorithm

Table 1. *Continued*

DV: Data Source and Style View of Diamonds(No model involvement except in DV-9)		
DV-1D	SQL/NoSQL/NewSQL?	Can add NoSQL sub-categories such as key-value, graph, document, column, triple store
DV-2D	Enterprise data model	e.g. warehouses. Property of data model highlighted in database community / industry benchmarks
DV-3D	Files or Objects?	Clear qualitative property of data model where files important in Science; objects in industry
DV-4D	File or Object System	HDFS/Lustre/GPFS. Note HDFS important in Apache stack but not much used in science
DV-5D	Archived or Batched or Streaming	Streaming is incremental update of datasets with new algorithms to achieve real-time response; Before data gets to compute system, there is often an initial data gathering phase which is characterized by a block size and timing. Block size varies from month (Remote Sensing, Seismic) to day (genomic) to seconds or lower (Real time control, streaming)
	Streaming Category S1)	S1) Set of independent events where precise time sequencing unimportant.
	Streaming Category S2)	S2) Time series of connected small events where time ordering important.
	Streaming Category S3)	S3) Set of independent large events where each event needs parallel processing with time sequencing not critical
	Streaming Category S4)	S4) Set of connected large events where each event needs parallel processing with time sequencing critical.
	Streaming Category S5)	S5) Stream of connected small or large events to be integrated in a complex way.
DV-6D	Shared and/or Dedicated and/or Transient and/or Permanent	Clear qualitative property of data whose importance is not well studied. Other characteristics maybe needed for auxiliary datasets and these could be interdisciplinary, implying nontrivial data movement/replication
DV-7D	Metadata and Provenance	Clear qualitative property but not for kernels as important aspect of data collection process
DV-8D	Internet of Things	Dominant source of commodity data in future. 24 to 50 Billion devices on Internet by 2020
DV-9	HPC Simulations generate Data	Important in science research especially at exascale
DV-10D	Geographic Information Systems	Geographical Information Systems provide attractive access to geospatial data

Table 1. *Continued*

Pr: Processing (runtime) View of Diamonds Useful for Big data and Big simulation		
Pr-1M	Micro-benchmarks	Important subset of small kernels
Pr-2M	Local Analytics or Informatics or Simulation	Executes on a single core or perhaps node and overlaps Pleasingly Parallel
Pr-3M	Global Analytics or Informatics or simulation	Requiring iterative programming models across multiple nodes of a parallel system
Pr-12M	Linear Algebra Kernels	Important property of some analytics
	Many important subclasses	Conjugate Gradient, Krylov, Arnoldi iterative subspace methods
		Full Matrix
		Structured and unstructured sparse matrix methods
Pr-13M	Graph Algorithms	Clear important class of algorithms often hard especially in parallel
Pr-14M	Visualization	Clearly important aspect of analysis in simulations and big data analyses
Pr-15M	Core Libraries	Functions of general value such as Sorting, Math functions, Hashing
Big Data Processing Diamonds		
Pr-4M	Base Data Statistics	Describes simple statistical averages needing simple MapReduce in problem architecture
Pr-5M	Recommender Engine	Clear type of big data machine learning of especial importance commercially
Pr-6M	Data Search/Query/Index	Clear important class of algorithms especially in commercial applications.
Pr-7M	Data Classification	Clear important class of big data algorithms
Pr-8M	Learning	Includes deep learning as category
Pr-9M	Optimization Methodology	Includes Machine Learning, Nonlinear Optimization, Least Squares, expectation maximization, Dynamic Programming, Linear/Quadratic Programming, Combinatorial Optimization
Pr-10M	Streaming Data Algorithms	Clear important class of algorithms associated with Internet of Things. Can be called DDDAS Dynamic Data-Driven Application Systems
Pr-11M	Data Alignment	Clear important class of algorithms as in BLAST to align genomic sequences
Simulation (Exascale) Processing Diamonds		
Pr-16M	Iterative PDE Solvers	Jacobi, Gauss Seidel etc.
Pr-17M	Multiscale Method?	Multigrid and other variable resolution approaches
Pr-18M	Spectral Methods	Fast Fourier Transform
Pr-19M	N-body Methods	Fast multipole, Barnes-Hut
Pr-20M	Particles and Fields	Particle in Cell
Pr-21M	Evolution of Discrete Systems	Electrical Grids, Chips, Biological Systems, Epidemiology. Needs ODE solvers
Pr-22M	Nature of Mesh if used	Structured, Unstructured, Adaptive

References

1. Big Data and Extreme-scale Computing (BDEC). http://www.exascale.org/bdec/. Accessed 29 Jan 2016
2. Data Science Curriculum: Indiana University Online Class: Big Data Open Source Software and Projects (2014). http://bigdataopensourceprojects.soic.indiana.edu/. Accessed 11 Dec 2014
3. DDDAS Dynamic Data-Driven Applications System Showcase. http://www.1dddas.org/. Accessed 22 July 2015
4. HPC-ABDS Kaleidoscope of over 350 Apache Big Data Stack and HPC Technologies. http://hpc-abds.org/kaleidoscope/
5. NSCI: Executive Order - creating a National Strategic Computing Initiative, 29 July 2015. https://www.whitehouse.gov/the-press-office/2015/07/29/executive-order-creating-national-strategic-computing-initiative
6. NIST Big Data Use Case & Requirements. V1.0 Final Version 2015, January 2016. http://bigdatawg.nist.gov/V1_output_docs.php
7. Apache Software Foundation: Apache Flink open source platform for distributed stream and batch data processing. https://flink.apache.org/. Accessed 16 Jan 2016
8. Asanovic, K., Bodik, R., Catanzaro, B.C., Gebis, J.J., Husbands, P., Keutzer, K., Patterson, D.A., Plishker, W.L., Shalf, J., Williams, S.W., et al.: The landscape of parallel computing research: a view from Berkeley. Tech. rep., UCB/EECS-2006-183, EECS Department, University of California, Berkeley (2006). http://www.eecs.berkeley.edu/Pubs/TechRpts/2006/EECS-2006-183.html
9. Bailey, D.H., Barszcz, E., Barton, J.T., Browning, D.S., Carter, R.L., Dagum, L., Fatoohi, R.A., Frederickson, P.O., Lasinski, T.A., Schreiber, R.S., et al.: The NAS parallel benchmarks. Int. J. High Perform. Comput. Appl. 5(3), 63–73 (1991)
10. Baru, C., Rabl, T.: Tutorial 4 "Big Data Benchmarking" at 2014 IEEE International Conference on Big Data (2014). http://cci.drexel.edu/bigdata/bigdata2014/tutorial.htm Accessed 2 Jan 2015
11. Baru, C.: BigData Top 100 List. http://www.bigdatatop.100.org/. Accessed Jan 2016
12. Bryant, R.E.: Data-Intensive Supercomputing: The case for DISC, 10 May 2007. http://www.cs.cmu.edu/bryant/pubdir/cmu-cs-07-128.pdf
13. Bryant, R.E.: Supercomputing & Big Data: A Convergence. https://www.nitrd.gov/nitrdgroups/images/5/5e/SC15panel_RandalBryant.pdf. Supercomputing (SC) 15 Panel- Supercomputing and Big Data: From Collision to Convergence Nov 18 2015 - Austin, Texas. https://www.nitrd.gov/apps/hecportal/index.php?title=Events#Supercomputing_.28SC.29_15_Panel
14. Coates, A., Huval, B., Wang, T., Wu, D., Catanzaro, B., Andrew, N.: Deep learning with COTS HPC systems. In: Proceedings of the 30th International Conference on Machine Learning, pp. 1337–1345 (2013)
15. Ekanayake, J., Li, H., Zhang, B., Gunarathne, T., Bae, S.H., Qiu, J., Fox, G.: Twister: a runtime for iterative mapreduce. In: Proceedings of the 19th ACM International Symposium on High Performance Distributed Computing, pp. 810–818. ACM (2010)
16. Ekanayake, J., Pallickara, S., Fox, G.: Mapreduce for data intensive scientific analyses. In: IEEE Fourth International Conference on eScience (eScience 2008), pp. 277–284. IEEE (2008)

17. Ekanayake, S., Kamburugamuve, S., Fox, G.: SPIDAL: high performance data analytics with Java and MPI on large multicore HPC clusters, Technical report, January 2016. http://dsc.soic.indiana.edu/publications/hpc2016-spidal-high-performance-submit-18-public.pdf

18. Fox, G., Jha, S., Qiu, J., Ekanazake, S., Luckow, A.: Towards a comprehensive set of big data benchmarks. In: Big Data and High Performance Computing, vol. 26, p. 47, February 2015. http://grids.ucs.indiana.edu/ptliupages/publications/OgreFacetsv9.pdf

19. Fox, G., Chang, W.: Big data use cases and requirements. In: 1st Big Data Interoperability Framework Workshop: Building Robust Big Data Ecosystem ISO/IEC JTC 1 Study Group on Big Data, pp. 18–21 (2014)

20. Fox, G., Qiu, J., Jha, S.: High performance high functionality big data software stack. In: Big Data and Extreme-scale Computing (BDEC) (2014). http://www.exascale.org/bdec/sites/www.exascale.org.bdec/files/whitepapers/fox.pdf

21. Fox, G.C., Jha, S., Qiu, J., Luckow, A.: Towards an understanding of facets and exemplars of big data applications. In: 20 Years of Beowulf: Workshop to Honor Thomas Sterling's 65th Birthday October, Annapolis 14 October 2014. http://dx.doi.org/10.1145/2737909.2737912

22. Fox, G.C., Jha, S., Qiu, J., Luckow, A.: Ogres: a systematic approach to big data benchmarks. In: Big Data and Extreme-scale, Computing (BDEC), pp. 29–30 (2015)

23. Fox, G.C., Qiu, J., Kamburugamuve, S., Jha, S., Luckow, A.: HPC-ABDS high performance computing enhanced apache big data stack. In: 2015 15th IEEE/ACM International Symposium on Cluster, Cloud and Grid Computing (CCGrid), pp. 1057–1066. IEEE (2015)

24. Iandola, F.N., Ashraf, K., Moskewicz, M.W., Keutzer, K.: FireCaffe: near-linear acceleration of deep neural network training on compute clusters. arXiv preprint arxiv:1511.00175 (2015)

25. Jha, S., Qiu, J., Luckow, A., Mantha, P., Fox, G.C.: A tale of two data-intensive paradigms: applications, abstractions, and architectures. In: 2014 IEEE International Congress on Big Data (BigData Congress), pp. 645–652. IEEE (2014)

26. Kamburugamuve, S., Ekanayake, S., Pathirage, M., Fox, G.: Towards high performance processing of streaming data in large data centers, Technical report (2016). http://dsc.soic.indiana.edu/publications/high_performance_processing_stream.pdf

27. National Research Council: Frontiers in Massive Data Analysis. The National Academies Press, Washington (2013)

28. Qiu, J., Jha, S., Luckow, A., Fox, G.C.: Towards HPC-ABDS: an initial high-performance big data stack. In: Building Robust Big Data Ecosystem ISO/IEC JTC 1 Study Group on Big Data, pp. 18–21 (2014). http://grids.ucs.indiana.edu/ptliupages/publications/nist-hpc-abds.pdf

29. Reed, D.A., Dongarra, J.: Exascale computing and big data. Commun. ACM **58**(7), 56–68 (2015)

30. Trader, T.: Toward a converged exascale-big data software stack, 28 January 2016. http://www.hpcwire.com/2016/01/28/toward-a-converged-software/-stack-for-extreme-scale-computing-and-big-data/

31. Van der Wijngaart, R.F., Sridharan, S., Lee, V.W.: Extending the BT NAS parallel benchmark to exascale computing. In: Proceedings of the International Conference on High Performance Computing, Networking, Storage and Analysis, p. 94. IEEE Computer Society Press (2012)

32. Zaharia, M., Chowdhury, M., Franklin, M.J., Shenker, S., Stoica, I.: Spark: cluster computing with working sets. In: Proceedings of the 2nd USENIX Conference on Hot Topics in Cloud Computing, vol. 10, p. 10 (2010)
33. Zhang, B., Peng, B., Qiu, J.: Parallel LDA through synchronized communication optimizations. Technical report (2015). http://dsc.soic.indiana.edu/publications/LDA_optimization_paper.pdf
34. Zhang, B., Ruan, Y., Qiu, J.: Harp: collective communication on hadoop. In: IEEE International Conference on Cloud Engineering (IC2E) Conference (2014)

New Benchmark Proposals

Benchmarking Fast-Data Platforms
for the *Aadhaar* Biometric Database

Yogesh Simmhan[(⊠)], Anshu Shukla, and Arun Verma

Department of Computational and Data Sciences,
Indian Institute of Science, Bangalore, India
simmhan@cds.iisc.ac.in, shukla@ssl.serc.iisc.in, arun.verma100@gmail.com

Abstract. *Aadhaar* is the world's largest biometric database with a billion records, being compiled as an identity platform to deliver social services to residents of India. *Aadhaar* processes streams of biometric data as residents are enrolled and updated. Besides ∼1 million enrollments and updates per day, up to 100 million daily biometric authentications are expected during delivery of various public services. These form critical Big Data applications, with large volumes and high velocity of data. Here, we propose a stream processing workload, based on the *Aadhaar* enrollment and Authentication applications, as a Big Data benchmark for distributed stream processing systems. We describe the application composition, and characterize their task latencies and selectivity, and data rate and size distributions, based on real observations. We also validate this benchmark on Apache Storm using synthetic streams and simulated application logic. This paper offers a unique glimpse into an *operational* national identity infrastructure, and proposes a benchmark for "fast data" platforms to support such eGovernance applications.

1 Introduction

The Unique Identification Authority of India (UIDAI) manages the national identity infrastructure for India, and provides a biometrics-based unique 12-digit identifier for each resident of India, called *Aadhaar* (which means *foundation*, in Sanskrit). *Aadhaar* was conceived as a means to identify the social services and entitlements that each resident is eligible for, and ensures transparent and accountable public service delivery by various government agencies.

The scope of UIDAI is itself narrow. It maintains a database of unique residents in India, with uniqueness guaranteed by their 10 fingerprints and iris scan; assigns a 12 digit *Aadhaar* number to the resident; and as an authentication service, validates if a given biometric matches a given *Aadhaar* number by checking its database. Other authorized government and private agencies can use this UIDAI authentication service to ensure that a specific person requesting service is who they are, and use their *Aadhaar* number as a primary key for determining service entitlements and delivery. However, to guarantee the privacy of residents, UIDAI does not permit a general lookup of the database based on just the biometrics, to locate an arbitrary person of interest.

T. Rabl et al. (Eds.): WBDB 2015, LNCS 10044, pp. 21–39, 2016.
DOI: 10.1007/978-3-319-49748-8_2

UIDAI holds the world's largest biometric repository. As of writing, it has enrolled 981 M of the 1,211 M residents of India in its database[1]. It continues to voluntarily register pending and newly eligible ones, at an operational cost of about USD 1 per person. It also currently performs up to 2 M authentications *per day* to support a few social services, and this is set to grow to 100 M per day as the use of *Aadhaar* becomes pervasive. The *Aadhaar* database is 5× larger than the next publicly-known biometric repository, the US Homeland Security's OBIM (VISIT) program, which stores 176 M records on fingerprints, and processes 40× fewer transactions at 88.73 M authentications *each year* (or 245 K *per day*), as of latest data available from 2014 [10].

Clearly, the *Aadhaar* repository offers a unique Big Data challenge, both in general and specially from the *public sector*. The present software architecture of UIDAI is based on contemporary open source enterprise solutions that are designed to scale-out on commodity hardware [3]. Specifically, it uses a Staged Event-Driven Architecture (SEDA) [12] that uses a publish-subscribe mechanism to coordinate the execution flow of logical application stages over batches of incoming requests. Within each stage, Big Data technologies *optimized for volume*, such as HBase and Solr, and even traditional relational databases like MySQL, are used. As part of the constant evolution of the UIDAI architecture, one of the goals is to reduce the latency time for enrollment of new residents or updation of their details, that takes between 3–30 days now, to something that can be done interactively. Another is to ensure the scalability of the authentication transactions as the requests grow to 100's of millions per day, and evaluate the applicability of emerging Big Data platforms to achieve the same.

Distributed stream processing systems such as Apache Storm and Apache Spark Streaming have gained traction, off late, in managing the data velocity. Such "fast data" systems offer dataflow or declarative composition semantics and process data at high input rates, with low latency, on distributed commodity clusters. In this context, this paper proposes benchmark workloads, motivated by realistic national-scale eGovernance applications, to evaluate the *quality of service* and the host *efficiency* that are provided by *Fast Data* platforms to process high-velocity data streams.

More generally, this paper highlights the growing importance of eGovernance workloads rather than just enterprise or scientific workloads [2]. As emerging economies like China, India and Brazil with large populations start to digitize their governance platforms and citizen-service delivery, massive online applications can pose unique challenges to Big Data platforms. For e.g., technology challenges with the HealthCare.gov insurance exchange website in the US to support the Affordable Care Act are well known[2]. There are inadequate benchmarks and workloads to help evaluate Big Data platforms for such public sector

---

[1] India 2011 Census, and live statistics from https://portal.uidai.gov.in.

[2] Healthcare.gov: CMS Has Taken Steps to Address Problems, but Needs to Further Implement Systems Development Best Practices, www.gao.gov/products/GAO-15-238.

applications. This work on characterizing applications for the identity infrastructure of the world's largest democracy is a step in addressing this gap.

The rest of the paper is organized as follows: in Sect. 2, we provide context for the UIDAI data processing workloads; in Sects. 3 and 4, we describe the composition of the *Aadhaar* enrollment and authentication dataflows, respectively. including characteristics of their input stream data sizes and rates; in Sect. 5, we evaluate this benchmark for Apache Storm using synthetic streams and tasks based on the real distributions; in Sect. 6, we review related work on Big Data benchmarks; in Sect. 7, we discuss how this benchmark can be expanded and generalized, and its relevance on the field; and finally offer our conclusions in Sect. 8.

2   Background

UIDAI aims to provide a standard, verifiable, non-repudiable identity for residents of India. This biometric-based identity distinguishes itself from other traditional forms of physical and digital identities in several ways. It is *unique, universal and non-repudiable*, which cannot be said for passports and driving licenses (not universal) or birth certificates and utility bills (not guaranteed to be authentic and unique). *Aadhaar* guarantees uniqueness of the individual by using their 10 fingerprints and iris scans of both eyes. It is also *electronically verifiable*, since it is just a number and the associated biometric of a person – it is not a "physical" identity card that can be lost, stolen or duplicated. Further, unlike digital identities like OAuth or OpenID, it also proves *physical presence* since the biometric has to be provided by the individual for authentication. Lastly, the fact that it has covered over 80% of eligible residents of India (5 years and older) makes it as close to universal as possible for a voluntary national identity program at this scale.

UIDAI offers two categories of services, one to enroll residents' demographic and biometric data into *Aadhaar*, and maintain them up to date; and another to authenticate users who provide their *Aadhaar* number and a biometric. We briefly offer context here, and in the next two sections, we drill down into the workloads themselves. Additional background can be found elsewhere [3].

2.1   Enrollment and Update

Residents who have not enrolled into *Aadhaar* register themselves with an authorized enrollment agency, and provide their *demographic details* (i.e., Name, Date of Birth, Address, EMail, Mobile Number) and their *biometrics* (photo, 10 fingerprints and both iris scans). This data is captured offline (presently), encrypted, digitally signed by the agency, and the *encrypted* enrollment packet for each resident uploaded daily to the UIDAI servers. As part of the enrollment pipeline, the resident's data is validated using basic sanity checks and, importantly, their biometrics is compared against every other resident's biometrics present in the database to ensure duplicate registrations are not performed. This de-duplication provides the authoritative uniqueness to *Aadhaar*. Once these checks pass, a random and unique 12 digit number is assigned and mailed to the resident. The packets remain encrypted right after data collection, stay encrypted on the network and on disk, and are decrypted in-memory, just in time, during data insertions, queries or comparisons.

The update process is similar, and allows residents who already have an *Aadhaar* number to change their transient demographic details, such as address and phone number, or even update their biometrics that could degrade with age.

2.2 Authentication and KYC

The basic *Authentication service* verifies if a given *Aadhaar* number matches a biometric or a demographic information provided by a resident, and returns a True/False response. This allows an authorized agency to request verification on whether the individual providing the biometric indeed matches the *Aadhaar* number they have on record, or if the demographic data that has been provided matches the verified one stored as part of the *Aadhaar* enrollment.

The service accepts the *Aadhaar* number and a combination of the following to be provided – fingerprint, iris scan, demographic (gender, age, etc.), One Time Personal Identification Number (OTP) – in an encrypted and signed manner. It then lookups up that *Aadhaar* number, matches the given biometrics and/or demographics against the details stored for that specific *Aadhaar* record, and returns a boolean response. The optional use of an OTP ensures that an authentication is done only with the consent of the individual who requests an OTP from UIDAI and provides it to the agency they are authorizing to perform their authentication.

Another *Know Your Client (KYC)* service extends this authentication to also allow the retrieval of demographic data and photo (but not fingerprint or iris scans) by an authorized agency, upon informed consent by the resident. This eases the burden of proof that residents have to provide in order to sign up for public or banking services, and promotes financial inclusion.

3 Enrollment Workload

In this section, we describe the proposed fast-data benchmark based on the *Aadhaar* enrollment application, followed by the authentication application in the next section. As part of the workload, we characterize the dataflow compositions used to process streaming data, the event rate and size distributions that they process, and the required quality of service in terms of end-to-end processing latency.

3.1 Enrollment Dataflow

A high level streaming dataflow composition for the Enrollment Workload is shown in Fig. 1. It also shows the *latency* per task to process an input packet,

Fig. 1. *Enrollment dataflow.* Tasks are labeled with the average latency time in milliseconds. The selectivity is given for each outgoing edge. "P" edges are taken by events that pass the check at a task, while "F" edges are taken by events that fail a check.

based on observations of the existing application logic, and its *selectivity*, i.e., the ratio between the number of items generated on an output edge for a certain number input items consumed.

The dataflow starts when encrypted enrollment packets (files) are uploaded to UIDAI by the registering agencies. The input packets' checksums have been verified at a DMZ to detect tampering, duplicate uploads, and malware/virus. The `Input` task emits such validated packets. Next, the `Packet Extraction` task decrypts the packets in-memory and inserts the demographics fields and the photo into a MySQL database and a Solr index, and the entire encrypted packet is archived into a distributed, geographically replicated store.

The `Demographic De-duplication` task locates existing residents whose demographics, such as name, age and pincode (zipcode), are similar to the incoming packet based on a Solr index search. Candidates that fuzzy-match the demographics have their biometrics preemptively checked against the input packet, and matching input packets, estimated at about 2%, are sent to the `Rejected` task. This avoids a full biometric de-duplication *across all residents* for these inputs.

Then, a `Quality Check` task does sanity checks on the demographics and the photo to ensure that the photo appears genuine (e.g., based on gender and age), the names and addresses appear valid, detects language transliteration errors during data entry, etc. An estimated 5% of enrollments are rejected by this task. Inputs that pass this QC arrive at the `Packet Validation` task where its digital signature is verified, and is checked to ensure it was generated by an operator who has been certified, is authorized to operate in that pincode, and matches their assigned supervisor and agency. Typically, 95% of packets pass this check.

Next, the `Biometric De-duplication` task performs a cross-check of each input packet's iris and fingerprints with every other registered biometric stored in *Aadhaar*. Three independent Automated Biometric Identification Systems (ABIS) are used here. The input biometric is first inserted into all of them, whereupon they extract a biometric template that they index. Then one of the ABIS is chosen, based on a weighting function, to verify if there is a match for this input from among all indexed biometrics. About 8% of inputs typically find a match and are flagged for potential rejection. Actual rejection of packets depends on `Additional Checks` of biometric and demographic data, a small number of which require manual checks, before being passed onto the `Rejected` task.

Packets that pass the prior tasks successfully are sent to the `Aadhaar-Generation` task which assigns a unique, random 12 digit number to this packet, creates a master record for the number, and forwards it for printing and mailing to the resident. Packets that are rejected in any prior step are retained by the `Rejected` task for auditing in a reject master database.

3.2   Enrollment Data Stream

The transaction rates for the enrollment dataflow change over time, and have to be captured to effectively evaluate the runtime performance of the application. Further, the input data sizes are larger than typical messages to stream processing systems, and hence their variation in sizes needs to be reflected too. Figure 2a shows the probability distribution function (PDF) for the encrypted input enrollment packet size that is uploaded by the enrollment agency and contains the demographics and biometrics. This distribution is from data collected on a single day in September 2015, and is representative of the overall trend. We see that half the packets have a 2–3 MB size, with the overall size ranging from 1–5 MB. We keep the input and output data sizes the same, for the purposes of the workload.

Next, Fig. 2b shows the average hourly input data rates within a recent 24 h period. This has a bi-modal distribution, with peaks in the late morning and late evening. The rate is low early in the day and by 11AM–12PM, ~65,000 packets/h are uploaded in batches by the field agencies after a morning session of registrations. Similarly, enrollment uploads from the afternoon session peaks at 6–7PM. Note that in 2013 when a bulk of the enrollments took place, the enrollment rate was much larger at 1.3 M packets per day. The output rate from individual bolts and the entire dataflow is determined by the selectivity.

In future, we expect the enrollment agencies to be constantly connected to the Internet, and the enrollment packet uploaded immediately upon capture. So, based on the latency achieved by the streaming enrollment pipeline, an *Aadhaar*

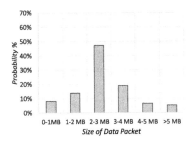

(a) Input Data Size Distribution

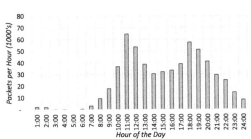

(b) Hourly Input Data Rate over a 24 hr day

Fig. 2. Probability distribution of the input data stream to the *enrollment dataflow*.

number can be assigned and returned interactively. The input rate distribution is expected to smoothen with such a model.

3.3 Expected Quality of Service

Each stream processing workload has a service level agreement (SLA) defined for it based on the end-to-end latency to process a single request. For the enrollment pipeline, the nominal throughput requirement based on a batch-processing model is to complete processing all packets that are received in a 24-hour period, within that 24-hour period. However, given that the advantage of a stream processing approach is to provide lower latency, we define the quality of service expected for processing each packet to be 10 min. This would allow residents to eventually enroll for *Aadhaar* interactively at a service center and be assigned an ID in the same session.

4 Authentication Workload

4.1 Authentication Dataflow

The authentication dataflow composition is shown in Fig. 3. This dataflow is pre-dominantly a linear pipeline with a selectivity of 1:1. As before, the latency is shown for each task based on an observational snapshot of the dataflow from Sep 2015.

The `Packet Validation` stage operates over HTTPS requests that arrive and is responsible for validating the authenticating agency and user, parsing the XML request and verifying the digital signatures of the request. Subsequently, the `Packet Decryption` task decrypts and extracts the packet's contents and verifies its integrity. The parameters of the request are then parsed by the `Verify Authorization` task, and a check performed on the type of authentication being done (demographic, biometric) and whether this request is a replay of a previous request. These tasks determine if the request is valid, and is being performed by an authorized entity.

After that, the *Aadhaar* number present in the request is used by the `Query Resident Data` task to lookup and retrieve the resident's demographic data from the backend HBase storage. The `Biometric and Demographic Match` task performs one or more of the following operations based on the type of authentication requested. It checks if an OTP or a PIN number, if present in the request, is valid. It may verify if the retrieved demographic matches the one in the request, if provided. And if a biometric is passed within the request, it checks if this biometric data matches the one stored for that *Aadhaar* number in the backend ABIS biometric system. These checks determine if the authentication failed or was successful.

As a measure of security, the `Resident Notification` task asynchronously notifies the owner of the *Aadhaar* number using their registered email or mobile number that an authentication was performed. Then an XML response is created with the results of the authentication and digitally signed by the `Create`

Fig. 3. *Authentication dataflow.* Tasks are labeled with the average latency time in milliseconds. Selectivity for all tasks is 1:1.

Response task. Finally, an audit record for the request is created, statistics on the authentication rates and latencies updated for business analytics, and the HTTPS response transmitted to the authenticating agency by the `Audit Log & Send` task.

4.2 Authentication Data Stream

The input stream to the authentication dataflow is presently more uniform than the enrollment dataflow, even as the number of users performing authentication will increase with time. Input requests are about 4 KB in size, and we use this constant size per message for our workload. The hourly input data rate is shown in Fig. 4. This is over 20× faster than the enrollment dataflow, and is also expected to grow in future. The current base rate for authentications is 150 requests/s, with two sharp peaks of about 500 requests/s in the morning and the evening – *Aadhaar* is used by federal employees to clock in and out of office each day, and these peaks reflect the requests by this attendance service[3]. In future, the number of authentications are expected to rise to 100 M requests during a working day, or an average of about 2,500 requests/s.

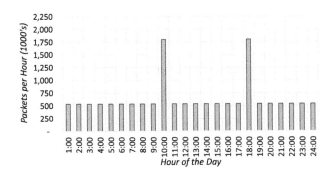

Fig. 4. Hourly input request rate over a 24 h day passed to the *authentication dataflow* with a base rate of 150 requests/s. The two peaks of 500 requests/s during the morning and evening periods reflect requests by the federal attendance service.

---

[3] Biometric Attendance Service, http://attendance.gov.in.

4.3 Quality of Service Expected

Authentication requests are inherently used by interactive applications. Hence the latency requirements for processing them are tighter. The SLA for end-to-end latency to process a single authentication request by the dataflow is set to 1000 ms, with most of the requests expected to be completed within 500 ms. In addition, there are network latencies for transmitting the request and response between the data center and the authenticating agency that add to the round trip time, which we do not consider here.

5 Experimental Validation

The motivation for the proposed workload is to evaluate the scalability of stream processing systems that can orchestrate the dataflow for the given inputs data streams, and return results within the specified SLAs. Reducing the computational resources required to achieve these quality of service metrics is an additional goal. We implement the proposed workload[4] and validate it on the Apache Storm distributed streaming platform.

5.1 Workload Generation

We compose the Enrollment and Authentication pipelines as two *topologies* in Apache Storm. Tasks are defined using a *synthetic bolt* logic that performs in-memory string operations for the given latency duration, and in the process consume CPU resources. The actual workload logic at UIDAI would have performed XML serialization or deserialization, encryption or decryption, remote NoSQL or a relational queries, and so on. Depending on the task, the load on the bolt itself would be limited to string parsing, integer arithmetic (both CPU bound) or waiting for query responses (idle CPU). So, in the absence of access to the actual logic themselves, the string processing performed by the synthetic bolt is a conservative approximation of the expected computational effort taken by each task in the dataflow.

The synthetic bolt is configured for each task to perform its computation for the given *latency* duration for that task. The bolt is also configured with the *selectivity* on each of the output edges for that task, and based on this, a certain fraction of input packets are routed to the downstream bolt(s) on the relevant edges. If the selectivity matches, the bolt logic passes the input packet to the relevant downstream bolt(s) without change.

We have developed an *event stream generator* tool that uses the given input rate and size distributions of the enrollment and authentication packets to pre-generate events with the appropriate relative timestamps and payload sizes. Synthetic events of these sizes are pre-fetched into memory by a *spout* logic within the Storm topology, and replayed at the appropriate relative time intervals to match the required input rate distribution. In particular, the spout uses a

---

[4] Code and data generator at https://github.com/dream-lab/bigdata-benchmarks.

multi-threaded and distributed mechanism to ensure that the data rate can be maintained at even 10,000's of events/s, if necessary. As we show, the results confirm that the observed input rates and sizes for the experiments match the reference input distribution that is given. The spout can also be configured with different time scaling factors to speed-up or slow-down the rates while maintaining a proportional inter-arrival time between events.

5.2   Deployment

We run the topology on a 24-node commodity cluster, with each node having 8 AMD Opteron 2.6 GHz Cores and 32 GB RAM, connected by GigaBit Ethernet, and running Apache Storm v0.9.4 on OpenJDK v1.7 and CentOS 7. Storm *supervisors* that execute the tasks of the topology run on 23 nodes and 1 node is dedicated to management services such as Master, Zookeeper and Nimbus. There are eight resource *slots* in each supervisor (host), one per CPU core, on which the Storm *workers* can execute one or more task threads.

Storm topologies can be configured to use only a subset of the supervisor slots in the cluster and also the degrees of parallelism (number of threads) assigned to each task in the topology. By default, each task runs on a single thread on a single slot. Based on the SLA required for each pipeline and data rate distribution, we determine the *minimum number of slots* that should be allocated to the topology and the *degree of parallelism per task* required to meet the SLA for the benchmarks. These configurations are based on the latency time per task in the dataflow such that there are adequate threads to process data arriving at the *peak rate* for that workload, and ensure that there are adequate CPU cores to sustain the compute requirements of these threads, yet without punitive context-switching overheads.

For the enrollment topology, the total degree of parallelism required to meet these latencies and peak rate comes out to 475 threads, shared uniformly across all the tasks, and running on 9 nodes (72 cores) of the cluster. For the authentication topology, we arrive at a total degree of parallelism of 514 threads, distributed proportionally across all tasks based on their contribution to the overall latency of the topology, and using 19 nodes (152 cores). The spouts that generate the input streams in parallel for the enrollment and authentication dataflows are included within this count and they take up 1 thread and 10 threads, respectively, while the terminal sink tasks in the topologies take up 1 thread each.

5.3   Results

Enrollment Workload. We perform a 24-hour benchmark run where data streams that follow the given rate and size distributions are generated and passed as input to the Enrollment topology in Storm for a whole day. Overall, about 591, 270 input packets were generated during this period.

Figures 5a and b show characteristics of the expected and actual input stream. We are able to validate that the event generator can generate input events with the same size distributions as the reference size distribution (Fig. 5a,

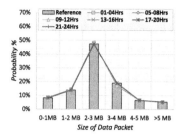
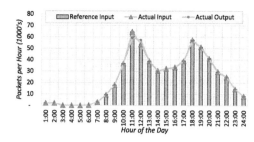

(a) Expected and actual input data sizes. Observed rates are grouped into 4-hour periods. We see that all periods have the same distribution.

(b) Expected hourly input rate (bar), and the actual hourly input and output rates (lines) over a 24 hr period.

Fig. 5. Probability distribution of the expected and observed input data stream to the *enrollment workload* experiment. (Color figure online)

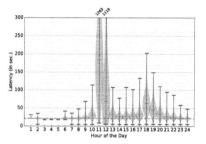
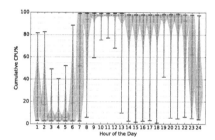

(a) Violin plot of latency distribution. The minimum theoretical latency for a successful enrollment is 21.22 secs, shown as a dashed green horizontal line.

(b) Violin plot of CPU utilization% sampled every second for 9 nodes, for a $1/10^{th}$ time duration experiment that is scaled back to 24 hours for plotting.

Fig. 6. Violin plot on hourly distribution of *latency* per packet (s) and *CPU utilization* for the *enrollment workload* experiment.

gray bars), not just cumulatively for the entire 24-hour period but also for individual 4-hour periods (Fig. 5a, lines). Likewise, we see from Fig. 5b that the hourly input rates generated by the spouts (green line/triangle) match the reference input rate distribution (gray bars). The figure also shows that the output event rate (orange line/circle) from the topology closely matches the input rates (green line/triangle). The output rate falls behind by ∼5,000 packets/h at 11 AM, where the input rate peaks, but it is able to compensate for this and catch up within the next two-hour period.

Drilling in further into the *event latencies*, Fig. 6a shows a violin plot[5] during each hour for the end-to-end latency for every event packet in that hour.

---

[5] The *violin plot* is a generalization of a box and whiskers plot. The minimum, median and maximum values are marked with a dash on the vertical line. The width of the horizontal shaded region around each vertical bar represents the relative frequency of packets having that latency value.

Here, we note that the average hourly *median* latency per enrollment packet is 35.74 s, relative to the theoretical lower bound makespan of 22.22 s per event that excludes the network time and other overheads. Except for the 11 AM and 12 PM hours, when we have a peak input rate and the system is catching up, the *maximum* observed latency time for packets during all other hours is less than 202 s per packet – well below the 600 s SLA.

For the 11 AM and 12 PM hours, the median latency is 183 s and 69 s, respectively, with a peak at 1,263 s. The evening input rate spike at 6 PM does not have such a strong impact on the latencies, and they go up only marginally to a median value of 62 s and a peak of 202 s. In all, 17,700 events out of a total of 591,270 events (i.e., 2.99%) had an SLA violation, where the end-to-end latency per event was greater than 600 s. In fact, *for 97% of the inputs, we are able to complete the enrollment pipeline within just 5 min.*

In order to evaluate the *resource efficiency* of Apache Storm in processing the workload, we sample the CPU utilization every second on each of the 9 active nodes for this topology. To keep the logs manageable, this is done for a shorter experiment that ran for $1/10^{th}$ the whole-day duration (i.e., 2 h and 24 min), while keeping the same size and rate distributions scaled down in time. Figure 6b shows an hourly violin plot of the CPU utilization% sampled every second per node, with each bar having $(9 \times 60 \times 60 \times \frac{1}{10})$ samples. As the pipeline warms up in the initial hour, we see that the CPU utilization is at about 20%, and beyond that, we see a correlation of the utilization with the input rate. Once we hit the morning peak corresponding to the 11 AM period, the CPU utilization is close to 100% for all nodes in the cluster during the entire hour, exhibiting high efficiency. The median utilization remains close to 100% for 13 straight periods, until the effects of the morning and evening peaks wear off by the 10 PM period. The average CPU utilization across all nodes for the entire period was high at 70%, despite the input rate variation.

Authentication Workload. The event generator for the authentication workload was also run for a 24-hour period and about 15,480,000 events are passed to the authentication topology during that time. Figure 5 shows that the reference and actual hourly input rates for the topology. The observed input rates are able to match precisely with the much higher, albeit smaller in size, expected input rates for the authentication workload compared to the enrollment workload. Since one output event is generated by each input event passed to the topology, the figure also shows that the output rate is able to sustain and match the input rate at the hourly granularity, even during the AM and PM spikes.

We examine the end-to-end event *latency* distributions for every hour using the violin plot shown in Fig. 8a. We see that except for the 10 AM and 6 PM hours that correspond to the input rate spikes, the bulk of the latencies for all other hours are tightly bounded, and fall predominantly in the range of 350–500 ms. This is in relation to the theoretical lower bound latency of 250 ms that does not take network costs into account. During the two spikes, the latency both increases and its distribution widens, with the most of the events falling in

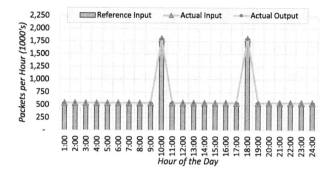

Fig. 7. Hourly reference and actual input data rates passed to the *authentication dataflow* during a 24 h experiment run, and the actual hourly output rates from the dataflow.

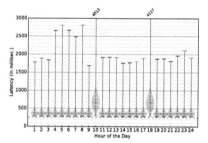
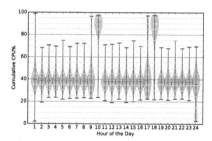

(a) Violin plot of latency distribution. The minimum latency for a successful authentication is 250 ms, shown as a dashed horizontal line.

(b) Violin plot of CPU utilization sampled every second for 19 nodes, for a $1/10^{th}$ time duration experiment that is scaled back to 24 hours for plotting.

Fig. 8. Violin plot of hourly distribution of *latency* per request (ms) and *CPU utilization* for the *authentication workload* experiment.

the latency range of 550–1,000 ms, but within the SLA of 1 s. While there are outliers in all hours that sometimes reach 4,500 ms, overall, we see that *the SLA is met for* 99.98% *of the input events*, and violated for only 2, 430 events out of over 15 M.

The violin plot of the *CPU utilization* across 19 nodes is shown in Fig. 8b, for a shorter experiment that ran for $1/10^{th}$ the period. Unlike the enrollment workload, we see that the median utilization stays low at about 40% for most of the hours except for the 10 AM and 6 PM input rate spikes when the median goes up to 93%. These utilization values however drop back in the next hour, indicating that the computation load stabilizes quickly after the input rate drops back to the base load of 150 requests/s. The average utilization across all hours and nodes is 43% (Fig. 7).

Relative to the enrollment workload, a higher fraction of input events fall well within their SLA limits for the authentication workload but we also observe a lower CPU utilization. This indicates that we can potentially trade-off resource efficiency and latency in the case of the authentication dataflow. The resource allocation could potentially be reduced for better efficiency, though the latency distribution for the two peaking hours may come closer to or go beyond the SLA limits. Similarly, the enrollment workload could improve its SLA beyond 97% by increasing the resource allocation, but consequently have a resource utilization lower than 70%. This also motivates the need for elastic resource allocation over time.

6   Related Work

Benchmarking distributed stream processing systems (DSPS) using real world workloads can help identify which of their features are impacted by unique input data characteristics, composition capabilities, and latency requirements. Several benchmarks have been proposed in this context.

The *Linear Road Benchmark* [1] simulates a highway toll system for motor vehicles with variable tolling, and was meant to compare Data Stream Management Systems (DSMS) with Database Management Systems (DBMS). The input to the benchmark is from a traffic model that has variable number of vehicles, but each emitting tuples at a uniform rate and with the same type. So while the input rate is variable, each message is of a fixed, small size. The metrics for evaluation are the response time and accuracy of the query, and the maximum sustained rate within a specified response time, but it does not consider resource utilization. While not developed for distributed stream processing systems, which deal with opaque messages and user logic, it can be adopted to validate DSPS. Our proposed benchmark is particularly tuned for DSPS and simulates the behavior of a real-world eGovernance workload, with variable message sizes and resource efficiency as an additional metric.

StreamBench [8] proposes 7 micro-benchmarks on 4 different synthetic workload suites generated from real time web logs and network traffic. Different workload suites are classified keeping performance metrics in mind. Performance workload and Multi-recipient performance workload suites measure the throughput and latency by pushing up the input rate, using single and multi receiver respectively; Fault tolerance workload suite measures the throughput penalty factor by deliberately causing nodes to fail; and Durability workload suite measures the fraction of time for which the framework is available when running the experiments for long durations. While we do not consider durability or fault-tolerance directly, our benchmarks are run for 24-hour, and this can be extended to longer periods to test durability. In addition to message rate variability, we also consider larger message sizes and variations in message sizes that can impact the resource utilization, which we additionally consider as a metric for comparison between DSPS platforms.

The *IBM Streams benchmark* [9] uses a email spam detection application over the Enron email dataset, and does a relative comparison of IBM's Infosphere Streams and Apache Storm. While they are able to reproduce the data sizes

(KBs) and rates from real emails from over a decade ago, their dataflow itself does not capture any application logic or even sleep times. So this benchmark is more a measure of the network overheads when running the dataflow, rather than computational resources used by, or the consequent latency of, the dataflow. Hence it has limited value.

While such stream processing benchmarks are useful, they do not consider message payloads that are large (MBs) in size, provide strict latency SLA requirement based on real applications, or capture characteristics of eGovernance services. Our *Aadhaar* benchmarks complements these efforts.

Chronos [5] is a generic data generation framework for stream benchmarks. Its focus is not to actually benchmark or stream the input data, but to generate data that can then be streamed to perform benchmarking. Chronos can create large scale synthetic time-series input data that mimics the distributions and correlations that it identifies from a given sample data. It uses Latent Dirichlet Allocation (LDA) method for extracting temporal dependencies from the sample data and preserves correlation among columns. This avoids users having to mine the data for complex patterns and distributions themselves for generating benchmark input specifications. This complements our work, where the distributions are explicitly given based on historic observations, and the focus is on benchmarking streams with different size and rate characteristics over the given dataflow definitions.

Other comprehensive benchmarks have developed for Big Data processing, beyond just fast-data processing which is our emphasis. *Hibench* [6] includes 10 micro workloads covering SQL Queries, Machine learning, Graph Computation and micro-benchmarks like word count and sort. *SparkBench* [7] comprises of different application types like Graph computation, SQL queries along with streaming application for Spark Streaming. This is a mixed-workload for different dimensions of Big Data. It uses resource consumption and data processing rate as the metrics for evaluation, but processing latency is not evaluated.

Big Bench [4] is modeled on the TPC-DS benchmark [11] and uses a retail industry scenario. It offers both semi-structured and unstructured data for data variety; analytics queries over click logs and user reviews, contributing to data volume; and an extract-transform-load (ETL) pipeline for data velocity. This is evolving as a community Big Data benchmark and has been validated on Teradata Aster and Hive platforms, among others. Variations to TPC-DS have also been proposed [13]. Similar to these Enterprise Big Data workload suites, one can envision an eGovernance Big Data workload suite, with this paper on the velocity dimension being a starting point.

7 Discussion

7.1 Extending the Workloads

The fast-data workloads that we have proposed are based on initial observations on the structure and latencies of the two *Aadhaar* dataflows, and the distributions of their input packet rates and sizes. There are several opportunities to expand upon these, both to increase the value derived from the workloads and to evaluate other Big Data platforms that support these applications.

The tasks in the two dataflows themselves are diverse in terms of the resources that they consume. Our use of a string processing synthetic task is a conservative proxy for the actual tasks. We can further *categorize the tasks in the dataflow based on their resource intensity* into CPU intensive, I/O intensive, memory intensive or idling tasks – the latter of which runs a remote query but does not consume much resources on the task's local host. Based on this, we can have different categories of simulated tasks be used in the workload which will help to model their resource consumption better. This will offer more genuine estimates of the degrees of parallelism required for each task, service level agreements that can be met, and the resource efficiency possible.

Occasionally, new and better biometric identification algorithms may be considered to improve the robustness of authentication. Then, all prior biometric data collected during enrollment may be *reprocessed* to evaluate the new algorithms. This reprocessing can use a variation of the enrollment pipeline to operate in a bulk mode, and compare the quality of the old algorithm with the new. There are also one-off scenarios when *bulk authentications* are performed for demographic verification by specific agencies, often during off-peak hours at night. Such high-throughput variations to the proposed low-latency workloads are worth considering.

The data rate distributions themselves are bound to change over time, and with *Aadhaar*'s use of multiple regional data centers, the load on the backend may have *region specific trends* that can be captured. It may also be possible to identify rate distributions specific to different authenticating sources, and generate synthetic *cumulative distributions* that blend them in different phases for representative, multi-modal, longer-term simulation runs. These can also help identify extreme spikes that may be possible when different distributions coincidentally peak at the same time, and help understand the behavior of the system during exigencies.

The trade-off that we observe between resource efficiency and SLA violations also provides the opportunity for developing and evaluating *resource allocation strategies* for fast-data platforms. The rate variation combined with the different probabilities for paths taken through the dataflow means that the load on each task is not uniform across time. Thus, these workloads can be used to benchmark the agility of fast-data platforms to intelligently acquire and release elastic Cloud resources to achieve the SLA while also reducing the (actual or notional) monetary cost for using virtualized resources.

While the stream processing dataflows coordinate the execution of the business logic, there are other high *data volume* workloads that are performed on the backend platforms that actually host the *Aadhaar* data and respond to queries from the streaming pipelines. These span NoSQL and relational databases that host demographic data, biometric databases that index and query over fingerprint and iris data, distributed file-systems that archive raw enrollment data and results, and audit logs from billions of authentications that are useful for mining.

Each of these are a Big Data platform case study in itself, and deserve attention as part of a eGovernance benchmark that cuts across Big Data dimensions.

7.2   Practical Considerations

The proposed *Aadhaar* benchmarks are validated using the Apache Storm distributed stream processing system. Other platforms such as Apache Spark Streaming and InfoSphere Streams could similarly be validated. Our results from evaluating Storm show that such fast-data platforms can achieve the SLAs that are required for the enrollment and authentication dataflows, and have the potential to significantly improve the quality of service for the end user.

However, such a validation of the orchestration platform is just one piece of a complex architecture, and cannot be construed to making an immediate operational impact within *Aadhaar*. The current SEDA model, which a distributed stream processing system could conceivably replace, is one of many Big Data platforms that work together to sustain the operations within UIDAI. Switching from a batch to a streaming architecture can have far-reaching impact, both positive and negative (e.g., on robustness, throughput), that needs to be understood, and the consequent architectural changes to other parts of the software stack validated. Making any design change in a complex, operational system supporting a billion residents each day is not trivial.

There are also logistical considerations since the overall application workflow relies on agents on the field and technology availability outside the data center. Some of the tasks in the dataflow, such as `Quality Check` in the enrollment dataflow, has human agents in the loop who complement automated quality checks. These have to be modeled better to guarantee the notional SLAs we observe, or there should be sufficient confidence in complete automation. Mobile field offices that collect enrollment data do not have Internet connectivity round-the-clock, since they may be at remote villages, and currently upload the data in batches each day. Moving to an interactive enrollment process necessitates constant Internet connectivity, which may be possible in the near future but requires additional resources and planning.

So, in summary, our proposed benchmarks are able to quantitatively verify the intuition that stream processing platforms will offer better SLAs that the current batch design, and offer a meaningful, real-world workload to verify fast-data platforms. But the caveats mentioned above in operationalizing such an architecture limit their immediate impact within *Aadhaar*.

8   Conclusions

In this article, we have proposed a Big Data benchmark for high velocity, *fast-data* applications based on an *eGovernance* workload. This addresses a gap in existing benchmarks that are based on web or enterprise applications, and are often volume-oriented.

Some characteristics of the workload stand out. The enrollment dataflow's input messages have a larger size, atypical of event streams for fast-data platforms that tend to be in the order of KBs in size. The bi-modal distribution of data streams for both the workloads is seen in event streams from other domains too, and is consistent with human activity patterns that are intrinsic to eGovernance platforms. The enrollment dataflow also has control-flow built-in, when packets fail validation and take a different path, and this impacts the probability with which different paths are taken. This is captured by the outgoing edge's selectivity. Both of these variations impact resource utilization, and motivate the need for elastic resource provisioning. Better resource allocation models for determining the degree of parallelism per task will also be valuable, and help meet the SLAs while conserving resources.

The *Aadhaar* workload is unique in its scale, being the largest of its kind in the world, but the characteristics of this workload can be seen in other public sector services such as the Department of Motor Vehicles or the Passport Office issuing IDs, and hence generalizable. Offering a benchmark based on these eGovernance workloads allows us to validate Big Data platforms for these socially important services, and offers the research and practitioner community additional transparency into the internal working of such mission-critical operations. While these results can also help improve UIDAI's Big Data architecture and their SLAs, the practical limitations discussed earlier stand.

Besides the many extensions to this work that was discussed earlier, it is worth examining other such public sector workloads to understand features are intrinsic to them, and set them apart from enterprise workloads. This will help design more effective Big Data solutions and potentially open up research opportunities.

Acknowledgments. We are grateful for inputs provided by Dr. Vivek Raghavan from UIDAI, and UIDAI's public reports in preparing this article. The views and opinions of authors expressed herein do not necessarily state or reflect those of the Government of India or any agency thereof, the UIDAI, nor any of their employees.

References

1. Arasu, A., Cherniack, M., Galvez, E., Maier, D., Maskey, A.S., Ryvkina, E., Stonebraker, M., Tibbetts, R.: Linear road: a stream data management benchmark. In: VLDB (2004)
2. Baru, C., Marcus, R., Chang, W. (eds.): Use cases from NIST big data requirements working group V1.0. Technical report M0180 v15, NIST (2013). http://bigdatawg.nist.gov
3. Dalwai, A. (ed.): Aadhaar technology and architecture: principles, design. best practices and key lessons. Technical report, Unique Identification Authority of India (UIDAI) (2014)
4. Ghazal, A., Rabl, T., Hu, M., Raab, F., Poess, M., Crolotte, A., Jacobsen, H.A.: BigBench: towards an industry standard benchmark for big data analytics. In: ACM SIGMOD (2013)

5. Gu, L., Zhou, M., Zhang, Z., Shan, M.C., Zhou, A., Winslett, M.: Chronos: an elastic parallel framework for stream benchmark generation and simulation. In: IEEE ICDE (2015)
6. Huang, S., Huang, J., Dai, J., Xie, T., Huang, B.: The HiBench benchmark suite: characterization of the mapreduce-based data analysis. In: Agrawal, D., Candan, K.S., Li, W.-S. (eds.) New Frontiers in Information and Software as Services. LNBIP, vol. 74, pp. 209–228. Springer, Heidelberg (2011). doi:10.1007/978-3-642-19294-4_9
7. Li, M., Tan, J., Wang, Y., Zhang, L., Salapura, V.: Sparkbench: a comprehensive benchmarking suite for in memory data analytic platform spark. In: ACM International Conference on Computing Frontiers (2015)
8. Lu, R., Wu, G., Xie, B., Hu, J.: Stream bench: towards benchmarking modern distributed stream computing frameworks. In: IEEE/ACM UCC, 2014 (2014)
9. Nabi, Z., Bouillet, E., Bainbridge, A., Thomas, C.: Of Streams and Storms. Technical report, IBM (2014). https://github.com/IBMStreams/benchmarks
10. Office of the Chief Financial Officer: Office of Biometric Identity Management Expenditure Plan: Fiscal Year 2015 Report to Congress. Technical report, Office of Biometric Identity Management, Homeland Security, United States (2015)
11. Poess, M., Smith, B., Kollar, L., Larson, P.: TPC-DS, taking decision support benchmarking to the next level. In: ACM International Conference on Management of Data (SIGMOD), pp. 582–587. ACM (2002)
12. Welsh, M., Culler, D., Brewer, E.: SEDA: an architecture for well-conditioned, scalable internet services. In: ACM SOSP (2001)
13. Zhao, J.-M., Wang, W.-S., Liu, X., Chen, Y.-F.: Big data benchmark - big DS. In: Rabl, T., Jacobsen, H.-A., Raghunath, N., Poess, M., Bhandarkar, M., Baru, C. (eds.) WBDB 2013. LNCS, vol. 8585, pp. 49–57. Springer, Heidelberg (2014). doi:10.1007/978-3-319-10596-3_5

Towards a General Array Database Benchmark: Measuring Storage Access

George Merticariu[1], Dimitar Misev[2], and Peter Baumann[1(✉)]

[1] Jacobs University Bremen, Campus Ring 1, 28759 Bremen, Germany
{g.merticariu,p.baumann}@jacobs-university.de
[2] rasdaman GmbH, Hans-Hermann-Sieling-Str. 17, 28759 Bremen, Germany
misev@rasdaman.com

Abstract. Array databases have set out to close an important gap in data management, as multi-dimensional arrays play a key role in science and engineering data and beyond. Even more, arrays regularly contribute to the "Big Data" deluge, such as satellite images, climate simulation output, medical image modalities, cosmological simulation data, and datacubes in statistics. Array databases have proven advantageous in flexible access to massive arrays, and an increasing number of research prototypes is emerging. With the advent of more implementations a systematic comparison becomes a worthwhile endeavor.

In this paper, we present a systematic benchmark of the storage access component of an Array DBMS. It is designed in a way that comparable results are produced regardless of any specific architecture and tuning. We apply this benchmark, which is available in the public domain, to three main proponents: rasdaman, SciQL, and SciDB. We present the benchmark and its design rationales, show the benchmark results, and comment on them.

1 Introduction

Array databases have set out to close a gap in the information categories supported by databases. Following support of sets through relational databases, hierarchies through XML databases, semantic nets through RDF/SPARQL, and general graphs through graph databases, next storage and retrieval for massive multi-dimensional arrays is of prime importance for manifold "Big" and relevant data such as sensor, image simulation output, and statistics data in science and engineering and beyond [3].

In fact, we claim that the inability of classic DBMSs to deal with the array part in scientific and engineering data – which forms a major part of those – is the reason for the unfortunate discrimination between data and metadata: Metadata are considered small, smart, and queryable – often, they are called "semantic" having such properties in mind. Data, conversely, tend to be large, "unstructured" from a database perspective, and not amenable to versatile retrieval. In today's quest for "Big Data Analytics" this divide is hampering access and analytics massively. Array databases represent an opportunity to cross this chasm and achieve uniform, integrated retrieval on science and engineering data.

© Springer International Publishing AG 2016
T. Rabl et al. (Eds.): WBDB 2015, LNCS 10044, pp. 40–67, 2016.
DOI: 10.1007/978-3-319-49748-8_3

Historically, arrays have developed in two separate directions, distinguished by the amount of values actually existing in the array. Dense arrays, such as images and weather forecasts, contain non-null values almost everywhere. Sparse datacubes, on the other hand, contain only relatively few actual values – typically, a one-digit percentage. Driven by the huge business case, sparse array have been addressed by Online Analytical Processing (OLAP) technology in the context of business data and have found their way into standard products of all major DBMS vendors. Two distinct technological approaches have been established. Relational OLAP (ROLAP) represents array cells as relational tuples relying on materialized views [21,42]; due to the sparsity, enumerating the few existing cells of the large datacube is a very effective compression method. Still, ROLAP suffers from the many joins necessary for addressing, in particular when it comes to the frequently occurring summarization ("roll-up") operations. Multi-dimensional OLAP (MOLAP) relies on specialized data structures and, to overcome the summarization bottleneck, all possible summaries as so-called prematerialized aggregates, thus often performing better than ROLAP and resulting in smaller database sizes depending on the data density, however lacking the scalability of ROLAP [13,40,42].

Dense arrays have been largely ignored by the database community; at most variable-length byte strings (BLOBs) have been offered. These, however, do not provide sufficient semantics and are not operationally supported on array level, hence, useless in practice. SQL:99 in [20] introduced 1-D arrays, but only with minimal operational support for selecting single elements and concatenating arrays. This has led to a severe impedance mismatch, a divide between the programming model and the database capabilities [19]. The massive arrays remain in files, with limited access functionality, usually just file download and rarely array subsetting, thus encouraging ad-hoc designed data models for querying them.

Array database management systems (ADBMSs) have a potential for overcoming this, although as of today they mainly find application on dense data, such as satellite imagery and weather forecasts in Earth sciences [9] and gene expression simulations in Life sciences [32]. Many multi-Terabyte installations are known, for example, of the rasdaman ADBMS as showin in Fig. 1 [7].

Given the increasing number of known array engine implementations the need arises for neutral, objective comparison of these systems. The only known general ADBMS benchmark, SS-DB [16], utilizes selected practical use cases, but does not provide a rigorous, systematic approach. Further, SS-DB only considers 1-D, 2-D, and 3-D arrays while science data in practice incur higher dimensions, such as 4-D x/y/z/t weather and ocean data and even 5-D climate data which have an additional time axis. We attempt to fill this gap by complementing SS-DB with a systematic benchmark ranging up to 6-D data sets, but configurable to even higher dimensions.

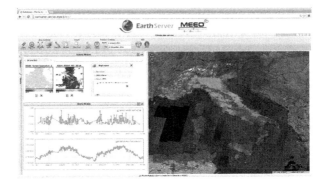

Fig. 1. Atmospheric timeseries analysis portal http://eodataservice.org/mea/ of the European Space Agency (ESA) powered by rasdaman, currently holding 130 TB and growing towards 1 PB.

In an earlier paper we have devised a strategy towards a comprehensive, systematic benchmark for ADBMSs [8]. In the paper on hand, we present one of the cornerstones of this roadmap, measuring storage access. The part of assessing ADBMS operations performance will be topic of a forthcoming paper. Given that arrays typically constitute "Big Data", the core critical functionality in storage management is extraction of sub-arrays, also called spatial subsetting[1]. Subsetting tests are systematically constructed in a way that does not allow systems to specifically tune themselves towards the tests performed. In particular, internal array partitioning is considered. Another reason for this approach is that some systems tend to implement array processing functionality in User-Defined Functions [16] which allows tailored handcrafted implementations of benchmark queries; by isolating storage access a more neutral assessment becomes possible. Goal of this paper, hence, is to selectively determine performance of the array storage manager.

We have applied the benchmark to three scientifically particularly important systems for verification: rasdaman, SciDB, and SciQL. However, this test applies to any ADBMS (as this functionality is at the heart of any such systems), and can even be applied to file-based solutions like OPeNDAP [14] which we plan to address in the near future as well. The benchmark code itself is publicly available [22] for take up, discussion, and evaluation of the same or further systems. This way we hope to establish, with the help of the research community, a standard analytical benchmark (complementing the domain-inspired SS-DB) to aid developers and service operators having to choose the system most suitable for their purpose.

The remainder of this paper is organized as follows. In the next section we briefly introduce ADBMS storage management techniques, followed by a state of the art review in Sect. 3. In Sect. 4 the ADBMS storage benchmark is described.

---

[1] "Spatial" in this context includes any axis, be it spatial, temporal, or of some other semantics in an application context.

The systems under test and conduction of the benchmark runs are introduced in Sect. 5, the results are presented and discussed in Sect. 6. Section 7 concludes the paper.

2   Storage Management in Array Databases

Arrays are commonly defined as a mapping a from some finite, non-empty d-dimensional integer interval D, i.e.: a subset of Euclidean space E^d, to some value space, functionally written as $a : D \rightarrow V$ [1]. The domain of an array is given by the cross product of the intervals defined by their lower and upper bounds, $D = I_1 x...x I_d$ for $1 \leq i \leq d$ where each interval has its individual integer lower or upper bound; note that bounds can go negative.

At each of the d-dimensional points of the function's domain (usually called *spatial domain* as discussed before) a *cell* is sitting, given by its coordinate vector and the value assigned to it. Normally, all array cells share the same *cell type*. For completely dense arrays a the mapping is total, otherwise it is partial. Non-existing values inside the array's domain are considered null, in practice often represented by some "don't-care value" such as 0, 255 or -9999.

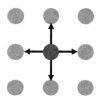

Fig. 2. Multi-dimensional Euclidean neighbourhood in arrays   (source: Wikipedia/ Peter Baumann).

The decisive difference between sets and arrays, which not only the conceptual model but also the storage and evaluation engine must align to, is that arrays have a well-defined Euclidean neighbourhood. Not only does this determine access behavior – when accessing the center pixel in Fig. 2 it is extremely likely that the surrounding pixels get accessed as well – but it also motivates the well-known efficient array storage structure where addresses are not materialized, but efficiently computed.

On principle, three options exist for storage of arrays: with coordinates suppressed, with coordinates materialized, and mixed approaches.

Alternative 1: Coordinate-free sequence. Cells are linearized following some axis order (like row major or column major order). By remembering the extent of the original array, cell positions can be computed. This is the storage approach for main memory arrays, classic (non-wavelet) image formats, and BLOBs. Storage is efficient as no coordinates need to be materialized. Access costs of this variant are mainly position and dimension dependent.

Consider the subsetting task sketched in Fig. 3 where the "x" marked subarray is to be extracted from the overall matrix. During storage linearization row clustering is preserved while column clustering is lost. Hence, along the innermost (last) axis burst read (as used in clumn store DBMSs) is fast while along all other axes the disk head has to move to different locations during reassembly, leading to disastrous overall performance. Additionally, if compression is applied then addressing capabilities are lost, and the – potentially large – array has to be expanded in main memory for subarray extraction.

Fig. 3. Serialization of an array for storage.

Alternative 2: Sequence independent, with explicit coordinates. Here, the array is represented as a set of coordinate/value pairs:

$$\{(x_1, f_1), (x_2, f_2), ..., (x_n, f_n)\}$$

Costs are not position correlated, but – despite index support – generally high as each single cell needs to be addressed separately. While this intuitively does not pay off for dense data it constitutes an efficient compression method on sparse data. In fact, this resembles ROLAP which considers sparse datacubes indeed.

Alternative 3: Partitioning, combined with sequencing within partitions. This two-level scheme (see Fig. 4) employs Alternative 1 on micro level; partitions are always accessed as a whole, hence the disadvantage mentioned above is not relevant. On macro level, Alternative 2 is used to maintain an inventory of the spatially located partitions; spatial indexing can help to speed up access by rapidly determining partitions addressed by a query through a range query. Access costs are low for bulk access of one partition, and usually not location correlated; only at partition borders neighbouring cells require access to another partition.

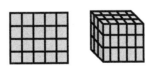

Fig. 4. Partitioned array storage.

Alternative 3 has been introduced by the imaging community at times when images became larger than main memory. ADBMSs commonly refer to such techniques as *tiling* [1] or *chunking* [36]. This method comes with several free parameters effectively acting as tuning parameters: indexing on macro level may speed up localizing the partitions needed for query answering, and a suitable partitioning can greatly reduce the number of partitions to be accessed. It is in particular the partitioning scheme which has critical impact on query response time. Hence, the more flexible the partitioning techniques of an ADBMS are the better it can be tuned towards subsetting workloads. In Sect. 3 we will inspect some ADBMSs on their capabilities in this respect.

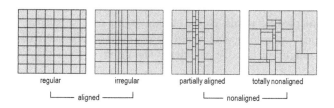

Fig. 5. Array partitioning schemes (after Furtado).

A comprehensive study of partitioning schemes has been undertaken by Furtado [18]. As shown in Fig. 5 she distinguishes between aligned and nonaligned partitionings, further subdivided into:

- *regular*, an aligned partitioning where all partitions have the same size and extent. This is particularly simple and efficient to implement as the home partition of a cell can be computed from its position. However, the only free parameter, partition size, allows only a very rough adjustment to query workloads.
- *irregular*, an aligned partitioning where still each axis has border coordinates common to all partitions at this point, but these border coordinates do not need to have equal distances between each other.
  From this scheme onwards some index is needed to look up partitions required for accessing some given array region.
- *partially aligned*, where along some axes the partition border coordinates "cut through" the whole array whereas along other axes partition borders are individual per partition.
- *non-aligned*, representing an arbitrary partitioning where the only constraint is that partitions remain arrays and the original array is given by the direct sum of its partitions. Obviously, this scheme allows the largest degree of freedom to adjust partition size, shape, and position to a given query workload.

ADBMS implementations use different storage schemes. SciQL [45] uses MonetDB's column store model for storing arrays which entails that multi-dimensional arrays get linearized as in Alternative 1 above. Most common, though, is Alternative 3. ArrayStore [37], the storage manager of SciDB, relies on it calling it *chunking*, rasdaman calls it *tiling* [5,18], and most further systems pursue this as well, as it is the generally accepted method to achieve data scalability. Variations exist in the spectrum of partitioning choices: while chunking relies on equi-sized partitions, tiling allows any possible partitioning into sub-arrays.

Souroush and Baladzinska [37] discuss an overlap scheme to accommodate focal functions, i.e., operations where a neighbourhood of an array cell is required to compute the new cell. This approach is commonly known in supercomputing as "halo" and used to speed up focal array access on supercomputer codes – see, e.g., [10]. Among the class of functions requiring such halo access are filter kernels and general convolutions. In non-redundant storage, evaluation of a focal function at partition borders requires access to neighbouring partitions which leads to undesirable multiple disk reads. By extending each partition with a "rim" of overlapping data focal operations can again be resolved locally on a single partition (as long as the neighbourhood stays within the limits of the rim).

3   Related Work

3.1   Array DBMSs

We present a very condensed overview on ADBMSs; see, for example, [6,35] for more comprehensive discussions and [2] for a minimal Array Algebra.

Vanilla implementations of ADBMSs reimplement the whole DBMS stack with components specifically tuned towards arrays. The rasdaman system [1,33] is a comprehensive from-scratch implementation, with all building blocks hand-crafted for arrays, from query frontend down to persistent storage management. EXTASCID [11,12] falls into this category, too.

Several implementations do not build an array engine from scratch (as rasdaman does), but rely on object-relational extension mechanisms. SciDB consists of a modified object-relational engine, a dedicated array storage manager, and array operations consistently implemented as UDFs based on ScaLAPACK [39]. SciQL [45] is an array query engine sitting on top of a conventional relational system maintaining arrays in a (linearizing)column store. 2-D geo imagery can be maintained in Oracle GeoRaster [29], however with very limited functionality. PostGIS Raster [28] supports 2-D and recently 3-D arrays. Tiles are recommended to be small (such as 100×100 pixels), and are visible to the query writer who has to recombine them explicitly into a result. Teradata, since version 13, has a feature called Array Support with some selected Array Algebra operations [41]. Size of one array is 64 kB minus size of an administrative infoblock, which seems to result in less than 50 kB actually available; concrete figures are not available from the manuals. Like PostGIS Raster single arrays need to be combined manually in queries.

A new project, TileDB [27], has been launched with the aim of allowing general partitioning like the rasdaman tiling, but constrained to sparse arrays. Not details are available currently.

Our tests include rasdaman, SciQL, and SciDB as the three currently most relevant available systems, as we will detail in Sect. 5.

3.2   Array DBMS Benchmarks

To the best of our knowledge, there is only one ADBMS performance benchmark published at the time of this writing, the Standard Science DBMS Benchmark

(SS-DB) [16]. The data used in this benchmark consists in astronomical images provided by Large Synoptic Survey Telescope (LSST) project, and varies in dimensionality and type from 1-D arrays (e.g., polygon boundaries), over dense and sparse 2-D arrays (image data and "cooked", i.e., processed data) up to 3-D arrays (trajectories in space and time). The proposed datasets are categorized into *small*, used for running the benchmark on a single machine and *normal*, used to run the benchmark on a distributed system composed of ten machines.

Data ingestion time is measured and it is part of the benchmarking procedure. For data processing and retrieval, there are three benchmarking categories for queries on raw data, queries on observations and queries on observation groups. Each category contains three queries. The queries apply operations based on custom algorithms, rather than standard array operations. In SciDB, the operations are implemented as user-defined functions (UDF) [16]. Thus, it is not clear whether SS-DB measures the performance of the array management system or the performance of the UDFs. In addition, SciDB is compared with MySQL, a relational database management system which lacks specific support for array processing. Consequently, it is not surprising that the performance of SciDB surpasses MySQL by orders of magnitude [35].

In [12] the SS-DB benchmark is used in order to compare the performance of two array database systems: SciDB and EXTASCID. The normal scale benchmark data was used, partitioning it and distributing it on 9 processing nodes. For execution time, the same query classes described in the SS-DB original paper were used. According to the presented results, EXTASCID performed on average 10 times better than SciDB when it comes to data loading and operations over "cooked" data since EXTASCID relies on extensive parallelisation of such tasks. However, SciDB performs better on raw data due to compression and chunk caching mechanisms.

While SS-DB is an important domain-specific benchmark its workload necessarily is a mix of different operations. Hence, it is not possible to selectively determine data load and processing - however, this is an important factor in particular as ADBMs are heavily CPU-bound as compared to classic, typically IO-bound DBMSs. The benchmark proposed establishes a complementary, domain-neutral, systematic testbed.

An ad.hoc performance evaluation of arrays in MongoDB can be found in [44]. While it only considers 1-D arrays up to about 10,000 elements aleardy it concludes "avoid large arrays" in MongoDB.

4   Storage Access Benchmark

In this section we present a domain-neutral, systematic benchmark for assessing ingestion and retrieval performance on multi-dimensional arrays. The core idea is to perform selection of sub-arrays and single values from n-D arrays. This constitutes one of the most fundamental operations and, therefore, has the potential for a high performance impact. To this end, subsetting operations are chosen which every known ADBMS supports.

The benchmark is heavily parametrized; key settings include:

1. maximum array dimensions, d, for evaluating 1-D through d-D arrays (default: $d = 6$);
2. maximum volume of the arrays queried (default: 1 TB);
3. maximum volume retrieved by the query workload (default: 1 GB).

This allows to run the benchmark code on platforms of different capabilities while ensuring comparability of results.

An important design goal was to determine access behavior independent from tuning towards particular access patterns. Queries are designed in a way that, based on our two decades of experience, the test cannot be outsmarted through an adjusted tuning – all queries work on the same datacubes which are exercised in manifold ways, as will be detailed below. In other words, any tuning will take effect in whole test suite execution.

Likewise, the test suite abstracts away from specific engine implementation. For example, resolving access requests through UDFs with customized code does not convey any particular advantage as there are no remarkable computations involved.

Access is measured on a single instance – parallelizing independent reads scales trivially with parallel data paths, such as in shared-nothing distribution, while the individual reads on each node will drive overall access performance. Hence, no particular insights can be gained from read parallelization measurements. On the other hand, efficiency of n architecture in accessing arrays as such can be measured, and there is a significant spread to be observed indeed as will be seen in Sect. 6.

The complete benchmark source code is publicly available on GitHub [22].

4.1   Test Data

As the benchmark is domain-independent it relies on synthetic data generated by a tool which is part of the benchmark code. By using random noise data we ensure that entropy is high enough that lossless comparison will not yield any advantage through smaller data – in other words, there is no benefit in data load times that could outweigh the decompression costs. The test checks that the originally ingested values are retained so that lossy compression will be flagged as returning wrong data.

Data size varies logarithmically over 1 KB, 100 KB, 1 MB, 100 MB, 1 GB, 10 GB, etc., per object. The increase in size of the datasets shows how a system scales up when size of the data goes up and also helps users to determine which system suits their needs depending on the performance factor within a data size range.

All arrays share the same base type *byte* (i.e., 8-bit unsigned integer) as the cell type structure does not impact data access times beyond the obvious increase of the total array and subset sizes. While rasdaman, for example, can handle even nested structs over heterogeneous base types we consider this not storage access in the narrow sens and, therefore, plan to consider such functionality in the forthcoming companion benchmark on array manipulation.

The domain extent of the arrays is cubic, that is: all edges share the same length which, for a d-dimensional array is obtained as the d-th root of the overall size imposed. Again, this does not constitute any restriction of generality as specific queries test query windows of varying edge ratio (see Sect. 4.2.3).

For each data size and dimensionality a randomly generated array is inserted into the tested DBMS, with system-specific storage parameters like partition size and index that theoretically seemed like most optimal. Ingestion time for each system is recorded and reported in Sect. 6.

4.2   Data Access

Data processing speed is influenced by multiple factors such as data localizations in the array, data read and applied operations. In the spirit of a step-by-step analysis, this paper is not concerned with measuring the performance of the applied operations and focuses only on data localization and retrieval.

Typically, in an array database the size of data reaches TBs in magnitude, while users normally retrieve only a small piece of an object, following the definition of "Big Data is data too big to be moved". Therefore, we define a parameter, MAX_SELECT_SIZE, which limits the maximum data size (in bytes) a test query will retrieve, regardless of the original array's dimensionality or size. Currently, this parameter is set to 10%, but can be modified by the tester depending on the test hardware's capabilities.

Five different patterns of data retrieval are tested in order to get as complete as possible picture of the storage manager's performance: *size queries* varying the size of the subarray extracted; *position queries* varying the location from where the subarray is taken; *shape queries* varying the edge ratio of the query box; *multiple select queries* to exercise non-local behaviour; and *center point queries* to inspect whether object size has any impact on retrieval time. Additionally, while building up the 1-D through 6-D test objects we measured ingestion time. We inspect each retrieval query set in turn.

4.2.1   Size Queries

This test provides information about how the retrieval time increases with the size of the data selected. Naturally, we would expect that access time grows proportionally to the size of the data retrieved. However, crossing partition boundaries and further implementation dependent characteristics may have an impact. In particular, when partitioning is involved there will be a stepwise increase of access times when crossing partition boundaries.

The test consists of a set of queries which all have the origin as lower bound, but grow their upper bound successively with a constant increment until MAX_SELECT_SIZE bytes are reached. Figure 6 shows the procedure schematically on a 2-D array. The selected size for each query is given by the following formula:

$$size = queryNO \cdot \left\lceil \frac{\texttt{MAX\_SELECT\_SIZE}}{6} \right\rceil \tag{1}$$

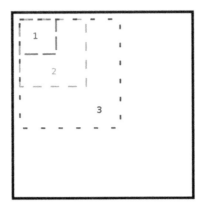

Fig. 6. Size queries: subsets incrementally growing in volume.

The reader may have noticed that the previous test (which focuses on sheer retrieval size) always has the lower corner of the retrieval window aligned with the bottom corner of the array. The question of what happens if this and other parameters are varied is going to be addressed with the next tests.

4.2.2 Position Queries

Purpose of this test set is to determine how access position affects system performance. The strategy adopted selects the minimum data size required, which is one point per partition loaded. This ensures that the partition access time is measured, and not the data transfer time. The maximum data requested is 64 bytes, which is reached with the 6-D array.

The method works as follows. First, the query selects a point from a partition. Then, two points are retrieved, one from each border of two neighbouring partitions. This goes on, until all neighbours are accessed. A requirement for this test case is that the partition overlap in any system is set to 0.

Figure 7 shows an example of two different selects over a 2-D array. The first query retrieves one value from a partition. Second query retrieves two points, each from two neighbouring partitions. Third query, retrieves the maximum number of neighbours for a partition.

A series of queries is defined where the query number, $QueryNO$, determines the number of partitions, p, that will be loaded according to the following formula:

$$p = 2^{QueryNO-1} \tag{2}$$

4.2.3 Shape Queries

Purpose of this test class is to determine how performance is impacted when the selected data shape varies. In particular, this shows behavior of the system under test when the access window does not match with any array partition shape.

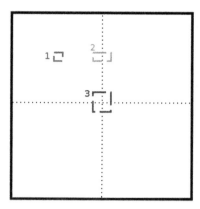

Fig. 7. Position queries: moving subsetting window across array.

Like in the other cases, the shape query test class only selects subsets of size MAX_SELECT_SIZE. However, unlike the previous two classes which select data with approximatively the same domain on each axis this class selects subarrays with varying axis extent ratios. Concretely, axis extents vary between 1 and MAX_SELECT_SIZE while keeping the query window at approximately the same volume, as much as arithmetically possible. Obviously, this test class does not apply to 1-D arrays. Figure 8 describes the shape selection over a 2-D array. The first two dimensions of the queries vary from a square to a rectangle with hight 1, always preserving the area of the shape. The domain shapes thus obtained are shown in Table 4 are determined by the following formulae:

$$dim = \left\lceil \sqrt[noOfDim]{\text{MAX\_SELECT\_SIZE}} \right\rceil$$

$$stepSize = \left\lceil \frac{dim}{6} \right\rceil$$

$$dim_0 = queryNO \cdot stepSize \tag{3}$$

$$dim_1 = \left\lceil \frac{\text{MAX\_SELECT\_SIZE}}{dim_0 \cdot dim^{(noOfDim-2)}} \right\rceil$$

$$queryDomain = [0 : dim_0, 0 : dim_1, 0 : dim, ...]$$

4.2.4   Multiple Select Queries

This class tests non-local access patterns where subsequent accesses refer to locations far apart along different dimensions.

The multiple select query class retrieves MAX_SELECT_SIZE bytes of data. However, the selection domains are split into two non-overlapping domains. The first window is fixed at the origin of the original array and the second window is translated along the diagonal of the original array. The diagonal direction was

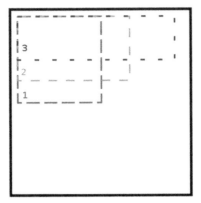

Fig. 8. Shape queries: extracting varying shapes from array.

chosen in order to avoid any axis-specific tuning. The second domain is computed as described in Eq. 4. Figure 9 shows sample consecutive selects over a 2-D array.

$$
\begin{aligned}
step &= \left\lceil \frac{\sqrt[noOfDim]{\frac{MAX\_SELECT\_SIZE}{2}}}{6} \right\rceil \\
lo &= queryNO \cdot step \\
hi &= (queryNO + 1) \cdot step - 1 \\
domain &= [lo : hi, lo : hi, ...]
\end{aligned}
\tag{4}
$$

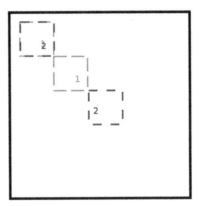

Fig. 9. Multiple select queries: spatially disparate extractions from array.

4.2.5   Center Point Queries

Purpose of this test is to determine whether the overall array size influences the data retrieval time. The test consist of a set of queries which retrieve only the center point of the array, again for test objects of multiple sizes. Only one cell is accessed so that delivery effort, once the cell is in main memory, is kept negligible.

5   Systems Under Test

The systems benchmarked – rasdaman, SciDB and SciQL – were installed on a standard Debian 7.0 system with an Intel Core i7-3770 K CPU (4×3.50 GHz, 8 MB L3 Cache), standard 1.8 TB 7200 RPM SATA hard disk, and 32 GB 1333 MHz RAM. Disk read/write speed was 134.5 MB/s and 137 MB/s, measured with the following commands respectively:

```
$ hdparm -t $device
$ dd bs=1M count=1024 if=/dev/zero of=o \
  conv=fdatasync
```

All tests were run sequentially on a single node; while all systems under test support distribution along multiple nodes, this benchmark targets the storage access performance which is unrelated to distribution of data.

Input data for the arrays were randomly generated with the Unix command

```
$ head -c size /dev/urandom
```

where the *size* parameter indicates the output volume in Bytes. This random file was then imported in the target system.

Partitioning capabilities were used[2] to establish a standard situation of regular partitioning with a partition volume of approximately 4MB. Earlier measurements have determined a size of few Megabytes as optimal on standard architectures [38]. The partition size for each dimension was set to $\lceil \sqrt[d]{t} \rceil$, with d as the dimension and t the total partition size in bytes. Although this sometimes created partitions of a size slightly larger than 4MB, the slight deviation was felt to be of no impact on measurements. Several models have been proposed for approximating an optimal partition shape given a set of subset queries [30, 36]. The partitioning in our benchmark generates symmetric partitions, however, in order to avoid query specific tunings.

Each query was executed five times. All results outside a 2σ[3] interval of the mean value got discarded and the remaining values averaged. For each query, the DBMS under test got restarted to achieve a cold database (i.e. do not allow the caching mechanism to influence the results); an exception was the "multiple queries" scenario where the query pairs described in Sect. 4.2.4 were executed in direct sequence before restarting the server. The repetition factor is a configurable parameter in the benchmark to allow for more or less statistical evidence. We found, though, that results were quite reproducible.

---

[2] Except in SciQL which does not support partitioning.
[3] Standard deviation.

5.1 Rasdaman

Historically, rasdaman is the first fully implemented Array DBMS with support for massive multi-dimensional arrays dating back to the mid-nineties [1], backed by a formal framework, Array Algebra [2]. For earlier compliance with the ODMG standard, arrays are stored in typed collections where each array has a unique OID. Hence, these collections can be considered column stores where each item is not a single value, but a whole n-D array of unlimited size. Currently, rasdaman is being integrated with SQL to implement the draft ISO Array SQL standard [23]. For array retrieval and processing rasdaman provides a declarative query language, rasql, which is based on SQL-92 [34]. UDFs add extensibility with external code, such as self-programmed C++ array operations or third-party tools like Hadoop and R.

Internally, arrays are partitioned in tiles which are managed directly in files or in database BLOBs [4,43]. While this partitioning remains transparent in retrieval, data designers have complete control over tiling and other physical parameters through a declarative storage layout language [5,18].

The random input data cube was inserted into rasdaman with the `rasql` client as binary arrays, by specifying its domain and cell type. In the benchmark, a standard aligned tiling in combination with an R+ tree index was adopted. For example, the 5-D array of size 1 MB can be inserted with the following query:

```
$ rasql -q 'create collection colD5S1MB ByteSet5D'
$ rasql -q 'insert into colD5S1MB values $1
            tiling aligned [0:20,0:20,0:20,0:20,0:20]
        -f randomData1MB.bin \
        --mddtype ByteArray5D
```

Select queries were similarly executed with `rasql`. The system was compiled from a fresh clone from the canonical repository [33], using version *rasdaman v9.2*. The benchmark uses the file store manager in rasdaman (i.e., no RDBMS like PostgreSQL for array storage). The server configuration file, `rasmgr.conf`, was adapted to start only a single server process, thereby emulating single-user mode. System restart after each query was performed by executing the standard commands:

```
$ stop_rasdaman.sh && start_rasdaman.sh
```

5.2 SciDB

SciDB [39] is another ADBMS supporting multi-dimensional arrays. A specialty is nested arrays, i.e., an array cell can be an array itself. This does not enhance expressiveness conceptually, though, as an n-D array with m-D sub-arrays in the cells can always be "flattened" to an $(n + m)$-D array. Arrays are broken into regular partitions, called *chunks*, which are stored on disk as units of access. Chunks have a fixed logical size, corresponding to regular tiling in Sect. 2. While this limits adaptivity to general access patterns chunking trivially reduces the addressing scheme so that no index is needed [39].

SciDB offers two languages for addressing the stored arrays. AFL (Array Functional Language) provides a set of User Defined Functions (UDFs) for loading, processing and retrieving array data. AQL (Array Query Language) is an adaptation of SQL for arrays. For this benchmark, AQL was chosen because it provides better flexibility for selecting data, especially for queries described in Sect. 4.2.4.

The randomly generated data is directly inserted in SciDB with the `iquery` client as binary file. The SciDB array domain is predefined and thus the file size must match the database array size. E.g. for a 5-D array of size 1 MB:

```
$ iquery -q "CREATE ARRAY colD5S1MB<ex:char> \
                    [axis0=0:15,21,0, \
                     axis1=0:15,21,0, \
                     axis2=0:15,21,0, \
                     axis3=0:15,21,0, \
                     axis4=0:15,21,0]"
$ iquery -q "LOAD colD5S1MB \
             FROM'randomData1MB.bin' \
             AS'(char)'"
```

Select queries were similarly executed with `iquery`. Following the manual [31], in order to get accurate results the *consume* function was used on the outputs of the SciDB select queries.

The system was installed using the `cluster_install.sh` script provided on the SciDB forum [26]. The configuration file used defines one server, with one instance per server, the number of CPU cores and the available RAM according with the values presented in Sect. 5. System restart is performed by executing `scidb.py startall <cluster_name>` and `scidb.py stopall <cluster_name>`.

5.3 SciQL

SciQL [45] is a query language that extends and enriches SQL with arrays as first class citizens. Its goal is to integrate array and set semantics, allowing indexed access of array cells via named dimensions, and making use of the windowing scheme introduced in SQL:2003 to group cells into so-called tiles (not to be confused with the rasdaman tiles), which can be further used to perform operations like statistical aggregation.

SciQL is implemented on top of MonetDB in a yet unreleased branch SciQL-2 available at [24]. MonetDB is a column-store database, and arrays in SciQL are mapped to columns in the database. So by including SciQL in this benchmark we can get an insight to an interesting question: is the column-store storage paradigm adequate for large multidimensional arrays?

The randomly generated data is transformed into a `COPY RECORDS INTO` statement, listing coordinates and value for each cell position on a new line in a single SQL file. Inserting a 5D array of size 1 MB was done with:

```
$ mclient -d benchmark < randomData1MB.sql
```

where `randomData1MB.sql` contains:

```
CREATE ARRAY colD5S1MB (
  axis0 INT DIMENSION [0:1:16],
  axis1 INT DIMENSION [0:1:16],
  axis2 INT DIMENSION [0:1:16],
  axis3 INT DIMENSION [0:1:16],
  axis4 INT DIMENSION [0:1:16],
  v TINYINT
);
COPY 1419857 RECORDS INTO "sys"."cold5s1mb"
    FROM stdin USING DELIMITERS ' ','\n','"';
0 0 0 0 0 -38
0 0 0 0 1 -5
0 0 0 0 2 -83
0 0 0 0 3 31
...
16 16 16 16 15 -74
16 16 16 16 16 75
```

Select queries were executed with the MonetDB JDBC driver. The system was compiled from source in a standard way:

```
$ ./configure --prefix=$SCIQL_HOME \
              --disable-strict --disable-jaql
```

A single database is created and used for the whole benchmark

```
$ monetdb create benchmark
$ monetdb release benchmark
```

Local database access is configured with setting `user` and `password` for the benchmark detabase in a `$HOME/.monetdb` configuration file. Restarting the server for each query execution is done as in the following pseudo code:

```
monetdbd stop $dbfarm
wait until $dbfarm/.merovingian\_lock is removed
monetdbd start $dbfarm
```

6   Results

In the concrete tests, we decided to go up to six dimensional arrays because this covers the dominating case of 1-D through 4-D spatio-temporal datacubes as they frequently occur in science, engineering, and beyond. This way, the benchmark does not reinvent wheels of medium-dimensional (roughly, 6-D through 12-D) data covered by OLAP benchmarks [15] and very high-dimensional feature spaces like in multimedia databases [25]. Sizings have been chosen to cover realistic data sizes, in our case: up to 1 GB extraction from up to 1 TB arrays. Just to reiterate, all these parameters are configurable in the benchmark code as published.

For each dimension we ingested datasets of size 1 KB, 100 KB, 1 MB, 100 MB, 1 GB, 10 GB and 1 TB. However, already after ingesting a 1-D array of size 1 GB, SciQL became too slow even with 32 GB of main memory, requiring more than 4–5 min just to restart and orders of magnitude more time to run queries compared to rasdaman and SciDB, and already using up 80+GB hard disk space. For SciDB inserting arrays of 1 TB size was not possible because the insert query execution failed because it requiring more memory than available on the system; we assume that the whole array has to be loaded into RAM, rather than the piecemeal alternative provided by rasdaman, there called "partial update". Hence, due to system limitations, the benchmark had to be restricted to arrays of maximum size 1 GB for SciQL and 10 GB for SciDB. For rasdaman, we went to the nominal array size chosen, 1 TB; this value is parametrized in the benchmark and can be increased at any time (in fact, actual database sizes for rasdaman exceed this by several orders of magnitude [7]).

On the other hand, inspecting download sizes up to a GB resembles practice in "Big Data" applications where data shipping is avoided as much as possible in favour of code shipping, so we feel this is not a limitation; if needed, downloads can easily be increased in the benchmark's parametrization.

6.1   Data Ingestion

Figure 10 presents the findings for the three systems for all data sizes as a comparison of ingestion times for 3-D. Again, execution time is plotted logarithmically. From the results we can conclude that ingestion time, as expected, increases with data size as well as with dimensionality.

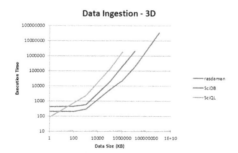

Fig. 10. Summary of ingestion times in ms.

Besides time, an important metric to note is the disk footprint occupied by the database (Table 1). SciQL in particular took a lot of disk space relative to the input data size, in contrast to SciDB and rasdaman which showed a storage footprint close to the input data size. Overhead in case of rasdaman is about 1.3%, in SciDB it is exceeding ten percent. This is surprising as SciDB, due to its regular partitioning scheme, does not involve indexes as rasdaman does to support irregular schemes.

Table 1. Disk space used for benchmark data.

System	Input data size	Disk space
rasdaman	66.66 GB	67 GB
rasdaman	6 TB	6.05 TB
SciDB	66.66 GB	74 GB
SciQL	6.6 GB	275 GB

6.2   Data Retrieval

We benchmarked the systems on several dimensions, with different array sizes, and various query workloads. In this section, we present all workloads and show representative results, together with a discussion of the behavior observed. For the purpose of this paper, to achieve some homogeneity, we uniformly use a retrieval size of 1 GB and 3-D arrays being close to the median of 1-D to 6-D. In the last subsection, we complement this walk-through by a discussion of cross-dimensional observations.

6.2.1   Size Queries

For this class six queries were run selecting an increasing percentage from the array, up to the full array. As a representative, we show the outcome for 1 GB 3-D datacubes in Fig. 11, represented as Query 1 to 6. Execution time is plotted logarithmically in milliseconds throughout the paper. The data size selected by each query is shown in Table 2.

Table 2. Data size selected for each query

Query number	Selected data size (MB)
1	170
2	340
3	510
4	680
5	850
6	1024

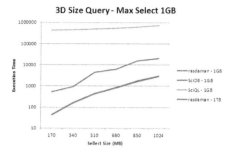

Fig. 11. Size query times in ms for 3-D arrays (array size: 1 GB, additionally 1 TB for rasdaman).

Naturally, execution times should increase with the selected data size, as we observe with rasdaman and SciDB. SciQL has a relatively constant time for the 1 GB case suggesting that the array is completely fetched into main memory before extraction is applied.

Interestingly, SciQL outperforms SciDB for the queries with higher indices, which roughly reflects the percentage of data extracted from the array. Maybe the strategy of SciQL to load the whole array and perform subsetting in RAM is superior when selections get closer to the full array. Of course, though, this does not scale to arrays larger than RAM (and might hint at why ingest into SciQL stopped early).

Figure 16 suggests that for systems using partitioning schemes (like rasdaman and SciDB) an increase in the array dimensionality will add a significant time penalty at the retrieval time. However, for SciQL, where no partitioning scheme is used, the overhead introduced by using more dimensions is not significant.

6.2.2 Position Queries

Four queries are executed in the 3-D array case in this query class. Table 3 lists the number of partitions loaded for each query.

Table 3. Number of partitions accessed by position query i.

Query number	Number of tiles
1	1
2	2
3	4
4	8

From Fig. 12 we can observe that SciQL and rasdaman performance is not affected too much by the number of loaded tiles when the data size is small; note that execution time is drawn logarithmically in milliseconds. SciDB's performance decreases significantly with the number of partitions loaded. For the

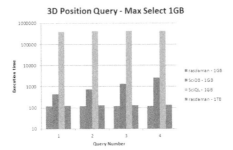

Fig. 12. Position query times in ms for 3-D arrays (array size: 1 GB, additionally 1 TB for rasdaman).

3-D example, the maximum data size selected was 8 bytes. Each byte was selected from a different partition, thus resulting in 8 partitions to be loaded. The results suggest that rasdamn's approach of indexing the partitions using a spatial index performs better than SciDB's approach of using same size partitions and no index. Also, using no partitioning scheme – the approach used by SciQL – comes with a substantial penalty compared with the other two systems.

6.2.3   Shape Queries

The tests consist of six queries selecting `MAX_SELECT_SIZE` data size. Query 1 starts with a "cubic" window which, over the subsequent queries, "flattens" into a long, thin window so that the box in Query 6 finally has one dimension of size 1 while always preserving overall data volume approximately. Table 4 presents the selection intervals for each dimension.

Table 4. Dimension upper bound for each query.

Query number	dim0	dim1	dim3
1	170	6165	1024
2	340	3084	1024
3	510	2056	1024
4	680	1542	1024
5	850	1234	1024
6	1024	1024	1024

Results in Fig. 13 suggest that for the "pure" array stores, rasdaman and SciDB, performance increases when the selection window spans wider along one dimension. Since the retrieval time pattern is the same for rasdaman and SciDB, the results suggest that rasdaman retrieval engine performs better than the SciDB one. Since SciQL does not allow any partitioning, overall performance is not affected from crossing partitions. However, compared with the other two systems, the execution time of SciQL for this type of queries is a few order of magnitude higher than for the other two systems.

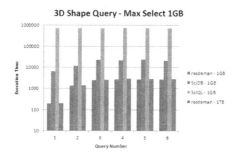

Fig. 13. Shape query times in ms for 3-D arrays (array size: 1 GB, additionally 1 TB for rasdaman).

6.2.4   Multiple Select Queries

In this class, a window of composite size `MAX_SELECT_SIZE` is selected in two non overlapping domains. All queries have the first domain fixed at 0, while the second domain moves along the diagonal of the original array. In case of Query 1, the second domain starts with the end of the first domain while in Query 6 the second domain has the same upper bound as the original array. Table 5 shows the selection domains for the second selection window.

Table 5. Lower and upper bound of each dimension for each query.

Query number	lo	hi
1	135	269
2	270	404
3	405	539
4	540	674
5	675	809
6	810	944

Performance is not affected by selecting different non-overlapping domains in the same query (Fig. 14). The retrieval time is constant from one query to another, suggesting that the partitions identification performs the same regardless the position of the selected index. Thus, the results boil down to which system has the best data retrieval engine. In this case, rasdaman outperforms both SciDB and SciQL. As always, execution time is a log 10 scale of *ms*.

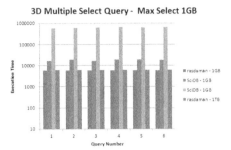

Fig. 14. Multiple select query times in ms (array size: 1 GB, additionally 1 TB for rasdaman).

6.2.5 Center Point Queries

This query class selects a single cell from the geometric center of object (or close by, in case of rounding). The result is always one byte, therefore its transport through the DBMS layers and to the client should have minimal impact on the response time.

The result was quite surprising. Naively, as SciQL with its columnar mapping does not employ partitioning its access times should be driven by the volume retrieved and remain independent from the underlying object size, hence, yield optimal values in this test. However, quite the contrary happened: Fig. 15 shows a significant performance degradation proportional to object size increase. This may be due to the fact that SciQL seems to be loading the whole array into main memory. On the other hand, rasdaman has the same performance regardless the object size. SciDB's performance is somewhat reduced, albeit not on small-size data, and overall relatively stable within the limits measured. This comparison suggests that indexing the partitions and loading only the necessary ones into memory (i.e., Alternative 3) constitutes a favourable strategy for large overall object sizes.

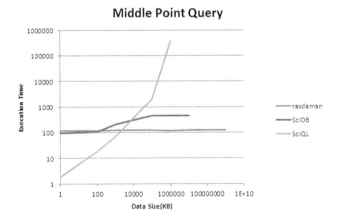

Fig. 15. Center point query times in ms.

6.2.6   Impact of Dimensions

In the last comparison, we look at the behavior of the systems under test depending on the array dimension, as another scalability aspect. Figure 16 is based on the same data from which the 3-D excerpts discussed earlier have been obtained, but measurements here are plotted against the number of dimensions.

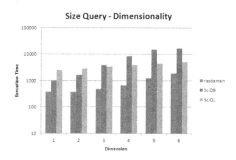

Fig. 16. Size query times in ms.

While there are differences in performance by about an order of magnitude there is a relatively weak dependency of each system on the number of dimensions. Again, we suspect that for SciQL this effect is based on the effect that the whole array is always read into main memory.

7   Conclusion

With the appearance of more and more Array Database systems, and Array Databases being an accepted research field, a domain-neutral comprehensive benchmark becomes an interesting endeavour. We are progressively establishing a benchmark for both storage access and processing on massive multi-dimensional arrays. This is intended to complement the domain-driven SS-DB benchmark towards reliable assessments of the young category of ADBMSs. Further, SS-DB compares an ADBMS (relying heavily on UDFs) with an RDBMS; hence, the benchmark does not easily generalize to an ADBMS comparison (where external code a la UDFs should be avoided).

In this paper, we have introduced the storage part of our growing array benchmarking suite. Operations are chose in a way that every ADBMS supports it, thereby ensuring comparability. A deliberately chosen slate of subsetting queries allows to determine the performance behaviour of an ADBMS under test, specifically designed so that the systems cannot unilaterally be tuned to a query workload. The test has been applied to three of the most important ADBMSs available, rasdaman, SciQL, and SciDB.

Results sometimes show convergence where architectures are similar, but occasionally also divergences of orders of magnitude in runtime performance. In particular, we can conclude that the column store paradigm (implementing

Alternative 1) does not scale, nor provide the same performance on large objects as partitioned storage (Alternative 3). For the systems which uniformly employ Alternative 3, rasdaman and SciDB, curves are overall similar although rasdaman is faster, except for very small arrays. This might be due to the fact that the rasdaman array engine has been crafted from scratch, with every single component optimized towards array handling in a many-years effort, whereas SciDB is using a combination of a Postgres kernel and UDFs which reduces implementation effort at the cost of less streamlined module interaction.

For the purpose of this discussion, only the most important results have been shown; however, results overall confirm findings as shown by the key measurements listed here. The full outcome will be made available as a Technical Report soon. Diagrams shown here focus on the sizings where a comparison was possible given the constraints of SciQL and SciDB. We have added 1 TB measurements for rasdaman which hint at smooth scalability: obviously, figures for 1 GB and 1 TB datacubes are essentially the same. Further, rasdaman appears two orders of magnitude faster than SciDB, which hints at quite some optimization potential for this young field.

To the best of our knowledge, this test is the first work available addressing a systematic evaluation of array engines. On the way to a comprehensive ADBMS benchmark there is a clear roadmap of further components which we will add step by step. This test has been designed for dense data in order to establish a clear platform for comparison, as outlined in Sect. 4. Of course, sparse data are relevant as well. Actually, there is no clear borderline, but rather a smooth transition between fully dense datasets (like meteorological simulation output), dense data with empty areas (such as cloud-cleaned satellite imagery and non-rectangular satellite imagery), and primarily empty spaces with isolated spots of payload data (such as statistical datacubes). Future work, therefore, will include adding tests for sparse arrays. To this end, test data sets of variable density will have to be designed which cover different distributions, such as homogeneous isolated data points versus large irregular empty areas. Among others, we want to investigate the effects of compressing both dense and sparse arrays on access performance.

The companion benchmark on operations will assess performance of operations on multi-dimensional arrays. As parallelization is a key scalability feature of any system when it comes to processing the benchmark likewise will have to test query splitting capabilities of the systems under test, including the efficiency achieved; rasdaman, for example, has proven successful in splitting multi-target queries over more than 1,000 cloud nodes with minimal parallelization overhead [17].

Finally, it will be interesting to subject more systems to testing. Further emerging ADBMSs, such as PostGIS Raster and EXTASCID, as well as non-DBMS array systems, like OPeNDAP and SciHadoop, should eventually be included as well for achieving a broad basis.

To further this, the whole test suite is provided in open source in a GitHub repository [22]. We encourage the community to participate in this endeavour by critically reviewing the tests and providing evaluations on further systems.

References

1. Baumann, P.: Management of multidimensional discrete data. VLDB J. **3**(4), 401–444 (1994)
2. Baumann, P.: A database array algebra for spatio-temporal data and beyond. In: Pinter, R.Y., Tsur, S. (eds.) NGITS 1999. LNCS, vol. 1649, pp. 76–93. Springer, Heidelberg (1999). doi:10.1007/3-540-48521-X_7
3. Baumann, P.: Array databases and raster data management. In: Oezsu, T., Liu, L. (eds.) Encyclopedia of Database Systems. Springer (2009)
4. Baumann, P., Dehmel, A., Furtado, P., Ritsch, R., Widmann, N.: The multidimensional database system rasdaman. In: ACM SIGMOD Record, vol. 27, pp. 575–577. ACM (1998)
5. Baumann, P., Feyzabadi, S., Jucovschi, C.: Putting pixels, in place: a storage layout language for scientific data. In: 2010 IEEE International Conference on Data Mining Workshops (ICDMW), pp. 194–201, December 2010
6. Baumann, P., Holsten, S.: A comparative analysis of array models for databases. In: Kim, T., Adeli, H., Cuzzocrea, A., Arslan, T., Zhang, Y., Ma, J., Chung, K., Mariyam, S., Song, X. (eds.) FGIT 2011. CCIS, vol. 258, pp. 80–89. Springer, Heidelberg (2011). doi:10.1007/978-3-642-27157-1_9
7. Baumann, P., Mazzetti, P., Ungar, J., Barbera, R., Barboni, D., Beccati, A., Bigagli, L., Boldrini, E., Bruno, R., Calanducci, A., Campalani, P., Clement, O., Dumitru, A., Grant, M., Herzig, P., Kakaletris, K., Laxton, L., Koltsida, P., Lipskoch, K., Mahdiraji, A., Mantovani, S., Merticariu, V., Messina, A., Misev, D., Natali, S., Nativi, S., Oosthoek, J., Passmore, J., Pappalardo, M., Rossi, A., Rundo, F., Sen, M., Sorbera, V., Sullivan, D., Torrisi, M., Trovato, L., Veratelli, M., Wagner, S.: Big data analytics for earth sciences: the earthserver approach. Int. J. Digit. Earth **9**, 3–29 (2015)
8. Baumann, P., Stamerjohanns, H.: Benchmarking large arrays in databases. In: Proceedings of the Workshop on Big Data Benchmarking, pp. 94–102, December 2012
9. Baumann, P., Yu, J., Misev, D., Lipskoch, K., Beccati, A., Campalani, P., Systems, G.I.: Preparing array analytics for the data Tsunami. In: Trends and Technologies. CRC Press (2014)
10. Benkner, S.: Hpf+: high performance fortran for advanced scientific and engineering applications. Future Gener. Comput. Syst. **15**(3), 381–391 (1999)
11. Cheng, Y., Rusu, F.: Astronomical data processing in EXTASCID. In: Proceedings of the 25th International Conference on Scientific, Statistical Database Management, pp. 47:1–47:4. ACM (2013)
12. Cheng, Y., Rusu, F.: Formal representation of the SS-DB benchmark and experimental evaluation in EXTASCID. Distributed and Parallel Databases, pp. 1–41 (2013)
13. Colliat, G.: OLAP, relational, and multidimensional database systems. SIGMOD Rec. **25**(3), 64–69 (1996)
14. Cornillon, P., Gallagher, J., Sgouros, T.: OPeNDAP: accessing data in a distributed, heterogeneous environment. Data Sci. J. **2**(5), 164–174 (2003)

15. T. P. Council. Tpc benchmark for decision support (tpc-ds). Accessed 31 Jan 2016
16. Cudre-Mauroux, P., Kimura, H., Lim, K.-T., Rogers, J., Madden, S., Stonebraker, M., Zdonik, S.B., Brown, P.G.: Ss-db: a standard science dbms benchmark. In: Proceedings of the XLDB Workshop (2010)
17. Dumitru, A., Merticariu, V., Baumann, P.: Exploring cloud opportunities from an array database perspective. In: Proceedings of the ACM SIGMOD Workshop on Data Analytics in the Cloud (DanaC 2014), pp. 1–4, 22–27 June 2014
18. Furtado, P., Baumann, P.: Storage of multidimensional arrays based on arbitrary tiling. In: Proceedings of the 15th International Conference on Data Engineering, pp. 480–489. IEEE (1999)
19. Gray, J., Liu, D.T., Nieto-Santisteban, M.A., Szalay, A.S., Heber, G., DeWitt, D.: Management in the coming decade. ACM SIGMOD Rec. **34**(4), 35–41 (2005). also as MSR-TR-2005-10
20. ISO. Information Technology - Database Language SQL. Standard No. ISO, IEC 9075: 1999, International Organization for Standardization (ISO) (1999)
21. Kimball, R., Ross, M.: The Data Warehouse Toolkit: The Complete Guide to Dimensional Modeling, 2nd edn. John Wiley & Sons Inc., New York (2002)
22. Merticariu, G., Misev, D., Baumann, P.: ADBMS Storage Benchmark Framework (2015). https://github.com/adbms-benchmark/storage. Accessed 31 Jan 2016
23. Misev, D., Baumann, P.: Extending the SQL array concept to support scientific analytics. In: Conference on Scientific and Statistical Database Management, SSDBM 2014, Aalborg, Denmark, June 2014
24. MonetDB. MonetDB branches (2015). http://dev.monetdb.org/hg/MonetDB/branches. Accessed 31 Jan 2016
25. Narasimhalu, A.D., Kankanhalli, M.S., Wu, J.: Benchmarking multimedia databases. Multimedia Tools Appl. **4**, 333–356 (1990)
26. n.n. SciDB. http://www.scidb.org/forum. Accessed 31 Jan 2016
27. n.n. Tiledb (2015). http://157.56.163.165/. Accessed 01 Jan 2016
28. Obe, R., Hsu, L.: PostGIS in Action. Manning Pubs. (2011)
29. Oracle. Oracle Database Online Documentation 12c Release 1 (12.1) - Spatial and Graph GeoRaster Developer's Guide (2014)
30. Otoo, E.J., Rotem, D., Seshadri, S.: Optimal chunking of large multidimensional arrays for data warehousing. In: Proceedings of the ACM 10th International Workshop on Data Warehousing and OLAP, DOLAP 2007, pp. 25–32. ACM, New York (2007)
31. Paradigm 4 Inc., SciDB Reference Manual: Community and Enterprise Editions, 2015. Accessed 31 Jan 2016
32. Pisarev, A., Poustelnikova, E., Samsonova, M., Baumann, P.: Mooshka: a system for the management of multidimensional gene expression data in situ. Inf. Syst. **28**(4), 269–285 (2003)
33. Rasdaman. The rasdaman Raster Array Database. http://rasdaman.org. Accessed 28 Feb 2015
34. Rasdaman. rasdaman Query Language Guide, 9.2nd edn. (2016)
35. Rusu, F., Cheng, Y: A survey on array storage, query languages, systems. arXiv preprint arXiv: 1302.0103 (2013)
36. Sarawagi, S., Stonebraker, M.: Efficient organization of large multidimensional arrays. In: Proceedings of the 10th International Conference on Data Engineering, pp. 328–336. IEEE Computer Society, Washington, DC (1994)
37. Soroush, E., Balazinska, M., Wang, D., ArrayStore: a storage manager for complex parallel array processing. In Proceedings of the ACM SIGMOD International Conference on Management of Data, SIGMOD 2011, pp. 253–264. ACM, New York (2011)

38. Stancu-Mara, S., Baumann, P.: A comparative benchmark of large objects in relational databases. In: Proceedings of the International Symposium on Database Engineering & #38; Applications, IDEAS 2008, pp. 277–284. ACM, New York (2008)
39. Stonebraker, M., Brown, P., Zhang, D., Becla, J.: SciDB: a database management system for applications with complex analytics. Comput. Sci. Eng. **15**(3), 54–62 (2013)
40. Szépkúti, I.: Multidimensional or Relational? How to Organize an On-line Analytical Processing Database. arXiv preprint arXiv: 1103.3863, March 2011
41. Teradata Corporation. Teradata Database, Tools and Utilities Release 13.10 (2013)
42. Vassiliadis, P., Sellis, T.: A survey of logical models for OLAP databases. SIGMOD Rec. **28**(4), 64–69 (1999)
43. Widmann, N., Baumann, P.: Efficient execution of operations in a DBMS for multidimensional arrays. In: Proceedings of the Tenth International Conference on Scientific and Statistical Database Management, pp. 155–165. IEEE (1998)
44. Wieruch, R.: Mongodb: Avoid large arrays - benchmark (2014). http://www.robinwieruch.de/avoid-large-arrays-in-mongodb-benchmark/. Accessed 31 Jan 2016
45. Zhang, Y., Kersten, M.L., Ivanova, M., Nes, N.: SciQL, bridging the gap between science and relational DBMS. In: IDEAS, pp. 124–133. ACM (2011)

Supporting Technology

ALOJA: A Benchmarking and Predictive Platform for Big Data Performance Analysis

Nicolas Poggi[✉], Josep Ll. Berral, and David Carrera

Barcelona Supercomputing Center (BSC),
Universitat Politcnica de Catalunya (UPC-BarcelonaTech), Barcelona, Spain
{nicolas.poggi,josep.berral,david.carrera}@bsc.es
http://aloja.bsc.es

Abstract. The main goals of the ALOJA research project from BSC-MSR, are to explore and automate the characterization of cost-effectiveness of Big Data deployments. The development of the project over its first year, has resulted in a open source benchmarking platform, an online public repository of results with over 42,000 Hadoop job runs, and web-based analytic tools to gather insights about system's cost-performance (ALOJA's Web application, tools, and sources available at http://aloja.bsc.es). This article describes the evolution of the project's focus and research lines from over a year of continuously benchmarking Hadoop under different configuration and deployments options, presents results, and discusses the motivation both technical and market-based of such changes. During this time, ALOJA's target has evolved from a previous low-level profiling of Hadoop runtime, passing through extensive benchmarking and evaluation of a large body of results via aggregation, to currently leveraging Predictive Analytics (PA) techniques. Modeling benchmark executions allow us to estimate the results of new or untested configurations or hardware set-ups automatically, by learning techniques from past observations saving in benchmarking time and costs.

1 Introduction

Hadoop and its derived technologies have become the most popular deployment frameworks for Big-Data processing, and its adoption still on the rise [15]. But even with such broad acceptance in industry and society, it is still a very complex system due to its distributed run-time environment, and its flexible configuration [7,10,17]. This makes Hadoop to be poorly efficient, and improving its efficiency requires knowledge of this complex system behavior. Not to mention all the emerging hardware and new technologies enhancing Hadoop that increase complexity of Hadoop systems.

The ALOJA project is a research initiative from the Barcelona Supercomputing Center (BSC) with support from Microsoft Research and product groups [11] to explore and produce a systematic study of Hadoop configuration and deployment options. The study includes the main software configurations that can greatly impact Hadoop's performance [7]; as well as different hardware choices

© Springer International Publishing AG 2016
T. Rabl et al. (Eds.): WBDB 2015, LNCS 10044, pp. 71–84, 2016.
DOI: 10.1007/978-3-319-49748-8_4

to evaluating their effectiveness and use cases including: current commodity hardware on which Hadoop systems where originally designed for [1], low-power devices, high-end servers, to new storage and networking[1]; as well as new managed Cloud services (PaaS). The intent of such study is to better understand how data processing components interact during execution and automate the characterization of Big Data applications and deployments. With the final purpose of producing knowledge and tools for the Big Data community, that can improve efficiency of already deployed clusters and guide the design of new-cost effective data intensive infrastructures.

This article presents the evolution of the project, initial results, and some lessons learned while benchmarking Hadoop for over a year continuously; iterating software configuration options and over a hundred different hardware cluster setups. At the same time, discusses the motivation both technical and market-based of such changes i.e., large search space and the emergence of economic Cloud services. As well as it overviews the different techniques employed to extract performance knowledge and insights from Hadoop executions, and exposes our current lines of research and focus. Since project beginnings almost two years ago, ALOJA's target and perspective has first evolved from a low-level profiling of Hadoop distributed environment, which allows to understand the networking bottlenecks and how components interact; to performing extensive benchmarking—which is still expanding, creating the largest public Hadoop performance repository so far with over 42,000 job executions. The repository is then used to evaluate via *aggregation* and *summarization* the performance and cost effectiveness of the different configurations available in the project's Web site, reducing both the size of data to be processed and stored. This reduction in data sizes allow us to share the platform a development virtual image to other researchers and practitioners directly, as well as to apply different Machine Learning (ML) techniques directly to the data [2].

While the results from the data *aggregation* efforts allows to process data interactively for the analytic online tools [2], the increasing number of configuration choices as the project expands in architectures and services—in the *millions* for benchmarks that a single iteration can take *hours* to execute, had led us to leverage Predictive Analytics (PA) techniques to be able to prioritize benchmarks and reduce the number of executions. PA encompasses a variety of statistical and ML techniques to make predictions of unknown events based on historical data [5, 13]—in this case the aggregated metadata of our benchmarking repository.

We use the *Intel HiBench* benchmark suite [18] as representative workloads, and we use the ALOJA-ML toolkit, the machine learning extension of the ALOJA framework [11] for such learning. The purpose of ALOJA-ML is to discover knowledge from Hadoop environments, first predicting execution times for given known workloads depending on the Hadoop configuration and the provided hardware resources; then evaluating which elements of a given deployment are the most relevant to reduce such running times. Our project goal is to find

---

[1] Storage PCIe NAND flash, SSD drives, and InfiniBand networking.

accurate models automatically to show and understand how our systems work, but also use them for making decisions on infrastructure set-ups, also recommendations for Hadoop final users towards platform and software configurations. At the same time supporting ALOJA's goal of automating Knowledge Discovery (KD) and recommendations to users.

An overview of the project's evolution summarized in Fig. 1, showing the different performance extraction techniques employed in the project, as well as the expansion to extract knowledge from Big Data applications to infrastructure providers. The next section describes in detail the motivation of such change in focus and efforts; while the pros and cons of each discussed with further details in Sect. 3.

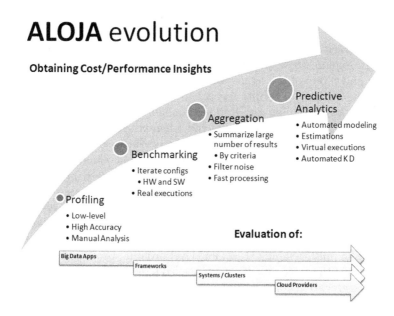

Fig. 1. Evolution of ALOJA: from performance profiling to PA benchmarking

2   ALOJA Evolution

During the development of the defined phases for project ALOJA [11], we have experienced a shift from an initial approach of using low-level HPC tools to profile Hadoop runtime [3] based BSC's previous expertise to higher-level performance analysis. Part of the initial work included inserting hooks into the Hadoop source code to capture application events, that are later post-processed into the format of BSC's HPC tools, which are used to analyze the performance and parallel efficiency of *supercomputing* or MPI-based workloads. However, due to the non-deterministic nature of Hadoop distributed execution, on top of the large number of software configuration options in Hadoop, some of the HPC tools were not directly applicable for finding best configurations for a particular hardware. Furthermore, our target has not been to directly improve the Hadoop framework codebase, but its deployment scenario.

To compare configuration options, during the first months of the project we started an extensive benchmarking effort on different cluster architectures and cloud services; iterating software configurations to extract execution times that were comparable. The resulting repository of benchmarks can be found at our web site [2]. As the number of benchmarks in the repository grew, evaluating internal results for each was no longer feasible, either for the low-level performance analysis tools or for manually revising them. Performance metric collection and log parsing both for profiling and benchmarking, have become Big Data problem in itself as results grow. For this reason, aggregation into summaries of the execution characteristics i.e., sums and averages. In this way turning execution details into meta-data, which allow us to contrast different results more efficiently; at the loss of information on execution internals, but reducing processing time. To explore these results more efficiently, we then developed different Web views and filters of the aggregated executions from the repository's online database. The analysis of results has led to a shift of focus in the project from the initial low-level profiling and Hadoop configuration, to cluster configuration and the cost/performance efficiency of the different systems.

Another reason for this change in perspective has been a shift from benchmarking on-premise to Cloud based clusters. Current cloud offerings for Big Data provide compelling economic reasons to migrate data processing to the Cloud [12] with the *pay-as-you-go* or even *pay-what-you-process* models. While the cloud has many benefits for cluster management i.e., dynamically scaling in servers, it also introduces challenges for moving data e.g., data is no longer local to nodes and it is accessed over the network. On the Cloud, due to the virtualized and public *multi-tenant* nature, low-level profiling becomes less relevant, as many samples of the same execution are needed to estimate average performance and results aggregation come into play. On top of this, for Platform-as-a-Service (PaaS) offerings, you might not have superuser access to the server to install packages or profiling the system [4].

Cloud providers also offer a great number of virtual machines (VM) options—at time of writing the Microsoft Azure Cloud offers over 32 different VM choices. Under this model, the same number of CPU processing cores and memory can be achieved by either having a larger number of small VMs (scale-out) or having a few larger VMs (scale-up) in a cluster. This great number of cluster configuration choices, which have an impact in the performance and the costs of executions [9, 14] has become one of the main targets of research and benchmarking efforts. Examples cloud VM size comparisons can found in our Web site [2]. These large number of Cloud topologies and services, hardware technologies, software configuration options that affect the execution time—and costs—of the different applications, leave us with millions of possible benchmark executions in the search space.

In order to cope with the increasing number of configuration options and new systems, the project was faced with the need first to do manual sampling from the search space, and grouping of results to extrapolate results between clusters. However, this initial approach has not been sufficient either, and still requires a large number of benchmarks. For this reason, we have leveraged Machine Learning (ML) techniques and implemented them as an extension of the platform. The generated prediction models allows us to estimate metrics such as job execution time for not-benchmarked configurations with great accuracy compared to classical statistical sampling and interpolation, as well as saving us time and execution costs. We are currently in the process of extending the use of such models on the platform to enable the Predictive Analytics (PA) [5,13] to complement the descriptive analytical tools available. PA techniques also complement our goal of automating Knowledge Discovery (KD) from the growing benchmark repository.

Having our efforts focused in PA, does not mean that we have stopped benchmarking or low-level profiling efforts. Each of the techniques have different uses cases and can complement each other i.e., PA is based on benchmarking metadata and profiling is used to debug or improve OS settings and configuration. An overview of the project's evolution summarized in Fig. 1, showing the different performance extraction techniques employed in the project, as well as the expansion to extract knowledge from Big Data applications to infrastructure providers. The next section compares each of the different performance extraction methods identified above.

3 Approaches to Extract Performance Knowledge from Hadoop

This section describes distinctive techniques to measure efficiency, performance, resource usage and costs employed by ALOJA.

3.1 Profiling

BSC having strong background with HPC workloads and their performance [3], the preliminary approaches to ALOJA consisted in instrumenting Hadoop, to make it compatible with well established performance tools used for parallel supercomputing jobs e.g., for MPI and related programming models. This initial work was achieved by developing the now *Hadoop Analysis Toolkit* which leverages the *Java Instrumentation Suite* (JIS) [6], a tracing environment for Java application, inserting hooks into the Hadoop source code; and a network sniffer based in *libpcap*. These changes allowed us to leverage the already existing low-level tools, simulators and knowhow from the HPC world, at the cost of having to patch and recompile Hadoop code[2]. Network traces allow us to study in great detail e.g., up to packet-level, the communication between components;

---

[2] Implementing dynamic code interposition is planned i.e., Aspect Oriented Programming.

both between nodes, as well as local data transfers as Hadoop uses remote and local services to transfer data.

While employing the *Hadoop Analysis Toolkit* for profiling, we are able to understand at specifically what timeframe certain Map/Reduce (M/R) steps i.e., shuffle, sort, merge, and spill. As well as how components synchronized in the cluster during the different distributed tasks. While profiling allow us a deep understanding of the underlying execution and debugging code, the level of detail can be daunting if not with the intent of optimizing the Hadoop code, or working with drivers or accelerators. Another problem that we faced with profiling, is the large amount of data that it produces. While the overhead to produce it is not a main concern, it leaves us with a Big Data problem in itself. Figure 2 compares the amount of data that would have been produced by profiling if enabled for the 42,000 executions in the repository. We currently enable low-level profiling only on selected traces, when a particular execution requires a deep level of analysis and debugging.

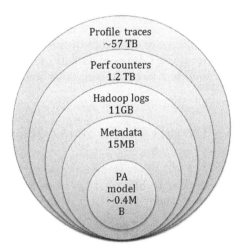

Fig. 2. Comparing the data sizes produced by each strategy (chart not in scale)

3.2   Benchmarking

Due to the large number of configuration options that have an effect on Hadoop's performance, to improve the efficiency of Hadoop requires manual, iterative and time consuming benchmarking and fine tuning over a myriad of software configuration options. Hadoop exposes over 100, interrelated configuration parameters that affect its performance, reliability and operation [7,9]. For this reason in ALOJA we have built an automated benchmarking platform to deal with defining cluster setups, server deployment, defining benchmarking execution plans, orchestration of configuration, and data management of results. While the platform is generic for benchmarking, our use case has been Hadoop.

Hadoop's distribution includes jobs that can be used to benchmark its performance, usually referred as *micro benchmarks*, however these type of benchmarks usually have limitation on their representativeness and variety. ALOJA currently features the HiBench open-source benchmark from Intel [8], which can be more realistic and comprehensive than the supplied example jobs in Hadoop. These are considered a representative proxy for benchmarking Hadoop applications, categorized into *micro-benchmark, search indexing, machine learning* and *analytical queries* types. We have currently gathered over 42,000 executions from the different HiBench benchmarks, that we use as base of our research to automate characterization and KD for Big Data environments.

In contrast to profiling, the benchmarking efforts operates more as black-box, where different Big Data frameworks and applications can be tested and summaries about their execution time and performance metrics are collected. However, we do include in ALOJA specific features for Hadoop, such as log parsers to detect the different phases in the M/R process, which could be extended for other frameworks if needed. Over time, collecting performance metrics have also become a Big Data problem: we have over 1.2 TB of performance metrics for the executions after importing the executions into the database. A description of the architecture in ALOJA can be found in [11]. While the data is 45x smaller than traces from profiling as can be seen in Fig. 2, as we get more executions, we currently use this data mostly for debugging executions manually. While we still keep it for every execution, summarizing data via aggregation has become more useful to extract cost and performance knowledge from groups of results.

3.3   Aggregation and Summaries

After a benchmark is executed and stored, the produced folder is then imported into a relational database. This process involves uncompressing data, transforming formats by using command line tools, parsing Hadoop logs, and importing the data into the database. This operation is a one time process, however the execution folders are kept in case they need to be reimported in the future due possible changes to the import routines or the use of external tools compatible with the logs. Once the data is on the relational database, ALOJA-WEB, the website component, interacts directly and interactively with it though SQL. This allows us very simply to use SQL *group by* statements on the data on different screens of the site.

As Hadoop job executions are non-deterministic, and several reasons can affect a particular execution performance i.e., multi-tenancy in the Cloud, or different random data generated by the benchmarks; all repeating executions are grouped and averaged (or other statistical functions are performed such as means or percentiles). Also, different views in ALOJA-WEB, allow to group together executions that share some common parameters i.e., same number of datanodes, disk configuration, or number of mappers; to complement missing executions and produce a common result. While these results need to be revised and validated, i.e., mixing executions with different cluster sizes might produce incoherent results; they can quickly offer a view on the tendency for the cost-performance. An example of these tools can be found in the *Config Evaluations* and the *Cost/Perf Evaluations* menus; including tools that allow to find the best configuration found with a particular filter, explore configuration parameters scalability, and rank clusters by cost-performance.

Aggregating results allows us to execute benchmarks only sampling the search space, and grouping to complement the results. Another feature is that it allows us to *average* noise or outlier executions automatically. While this is an improvement to analyzing benchmark result individually, it has several drawbacks: it exposes us to the problems of averages, that might not represent a well distributed value; also might lead to wrong conclusions if the grouped data is not analyzed and validated manually carefully, as mixing technologies and cluster sizes in these groups. For these reasons, and continue our goal towards automation, we have started the PA extension to overcome some of these shortcomings and automate Knowledge Discovery.

3.4 Predictive Analytics

While the results from the data *aggregation* efforts allows to process data interactively for the analytic online tools [11], the increasing number of configuration choices as the project expands in architectures and services—in the *millions* for benchmarks that a single iteration can take *hours* to execute. In order to cope with the increasing number of configuration options the project was faced with the need first to do manual sampling from the search space, and grouping of results to extrapolate results between clusters. However, this initial approach has not been sufficient either, and still requires a large number of benchmarks. For this reason, we have leveraged Predictive Analytics (PA) techniques to be able to optimally execute a smaller number of benchmarks. PA encompasses a variety of statistical and ML techniques to make predictions of unknown events based on historical data [5,13]—in this case the aggregated metadata of our benchmarking repository.

At this time, the ALOJA dataset has some challenging issues we have to deal with, in order to create as much representative models as possible. First of all, not all benchmarks have the same number of executions nor all the same executed configurations. Half of the ALOJA data-set examples are from *terasort* executions, while we can consider *pagerank* under-represented because of having much fewer executions. Executions use resources and time (thus they cost money), and the examples composing the data-set were executed following different reasons than to obtain a homogeneously representation of every benchmark and feature. One of the goals of ALOJA-ML is to be able to infer models from a not-so-regular data-set. In addition, the data-set contains failed or outlier executions, where external factors heavily affected the final execution time of the run. Some of those examples can be easily detected and removed, while others can not. Further, having a huge space of feature combinations, compared to the number of examples, can bring uncertainty when deciding if an example is an outlier or is a legitimate example lonely in its space region. At this time we will use the data-set without filtering, as a first approach to it without *massaging* our data.

Modeling Benchmark Behaviors. To model a system some examples of this system are collected, this is what we want to be able to predict from this system, and any element that can determine or affect it. The model is then a function that receives as inputs variables like in our case the benchmark name, the introduced Hadoop configuration and the hardware specifications, and produces as output the required execution time for the benchmark in such conditions. Then from the collected set of examples, the data-set, machine learning algorithms attempt to infer an explanation for these outputs from the provided inputs

Our current main focus is to obtain a model to predict benchmark execution times from a given software and hardware configurations, in the more automatic possible way. As previously said, Hadoop executions can be tuned through software configurations (number of mapping agents, type of compression used, etc.) and through the provided hardware resources. These tuning parameters determine the resulting execution, and we consider them our input variables. Also, as we focus on the resulting execution time, we consider that as out output variable we want to be able to predict. Being able to predict such values will let us to plan benchmark executions, compare different environments without even executing such benchmarks, also plan new data-centers by deciding their components from a variety of available hardware configurations while predicting the execution time of reference benchmarks with each of them. Figure 3 summarizes the machine learning schema, also applied to our case of study.

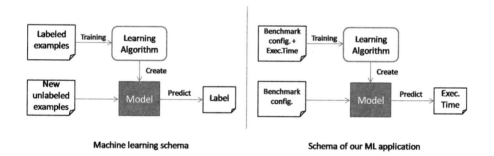

Machine learning schema   Schema of our ML application

Fig. 3. Machine learning methodology in our environment

4   Modeling Behaviors

4.1   Learning Models

Having different benchmarks available, and willing to cover with one model as much space of configuration possibilities as possible, raises our first concern, that is to check if they are comparable and in which degree, in order to model all of them in a generalist model. Each benchmark has different characteristics that led to different behaviors (CPU bound workloads, IO bound workloads, etc.). One could hope that training a model with samples of different benchmark executions, when including the identifier of each benchmark, the model would recognize this as an important feature and return different sub-models for each group of behave-alike benchmarks. One of the studies done is to check whether training a model with all the benchmarks produce an acceptable model against creating a model for each specific benchmark. In one hand, a generalist model requires one single training process, benchmarks with many examples in the data-set can complement other behave-alike ones with less examples. On the other hand, a specific model can fit better a benchmark, being aware of not over-fitting, but it requires enough examples of each one, and for each benchmark it requires a new training process.

For a first learning attempt, we selected as a machine learning algorithm the M5P, which creates a regression tree model [19, 20]. Learning is done by using a random split of 50% of the available data-set, 25% used to validate the parameters of the obtained model, and 25% kept for testing only. The selected variables input for the following experiments are *benchmark, network, storage type, number of maps, block size, cloud provider + kind of deployment, sort factor, file buffer size*, also *execution time* as output variable.

After learning models on each of the 8 tested benchmarks, we obtain different results for each one. Benchmarks like *bayes, kmeans, sort, terasort* and *wordcount* reach Relative Absolute Errors (RAE) between 0.11 to 0.23 during the testing process. This is that we can predict execution times of new executions with an expected bias between 11 and 23%. In the other hand, the *pagerank* benchmark has shown unstable executions, and for some executions their execution time in extremely high, making difficult to learn and predict (obtaining an unacceptable RAE of 1.18 in testing). The lighter side of this is that, from this process, we detected that such executions deserve a human review to see whether *pagerank* is actually unstable or something unexpected happened on our system when running some Hadoop jobs. Finally we observe how benchmarks *dfsioe_read* and *dfsioe_write* have high variance on their execution times, and in both training/validation and testing errors are high enough to re-think how to deal with them (RAE between 0.38 and 0.44 in testing). The fact that shared resources like disks are heavily involved in such benchmarks can produce this variance, so in the future modeling them should include variables representing the shared usage of such resources.

Once seen that most of the benchmarks can be modeled automatically with a tolerable error, we proceed to create a generalist model for the complete set. We had hope that the generalist model would do worse but not much than the specific ones. We obtain RAEs for all our benchmarks around 0.44 on training/validation and test, modeling all 8 benchmarks together also using all but the *dfsioes* and *pagerank* just to check whether the high error was caused by the hard-to-model benchmarks. This brings us to the conclusion that we should attempt to create separated models per benchmark, but it would be interesting to model *any* workload despite being benchmarks or not. A general model would help us to introduce a new workload, represented by any of the modeled benchmarks, and predict its execution time. Our future approaches focus on work-around the problem of being unable to generalize, by parametrizing benchmarks so any benchmark (even any properly parametrized workload) can be go through the same model, and trying to find the causes that make some benchmarks harder to predict or find proper new input variables that define better those benchmarks.

4.2   Representing Characteristics

Far away from just model benchmarks to do predictive analysis, we can use the obtained models to discover how each input affects the output variable. Instead of producing more executions or rely on the example executions we complete, for each configuration not tested (or even tested), the space of possible configurations with each expected execution time. For practical reasons we do this given a reduced set of input variables we can consider relevant, fixing the rest of inputs to a given value. Note that using all inputs could lead to millions of predictions, hence we select the inputs we consider more interesting to explore. With the resulting predictions we can rank each configuration from slower to faster, and observe how changing variables affect the result.

For this we use a greedy algorithm that separates the ranked configurations in a dichotomous way (Algorithm 1), finding which variables produce the biggest

---

Algorithm 1. Least Splits Algorithm

1: **function** LEAST.SPLITS(e)   ▷ e set of $\langle conf, pred \rangle$ ordered by $pred$
2:      **if** $|e| > 1$ **then**
3:          $bv \leftarrow null$; $lc \leftarrow \infty$
4:          **for** $i \in variables(e)$ **do**   ▷ Search variable with less changes
5:              $c \leftarrow 0$
6:              **for** $j \in [2, |e|]$ **do**
7:                  **if** $e[i, j] \neq e[i, j-1]$ **then**
8:                      $c \leftarrow c + 1$
9:                  **end if**
10:             **end for**
11:             **if** $c < lc$ **then**
12:                 $\langle bv, lc \rangle \leftarrow \langle i, c \rangle$ ▷ Variable i is candidate
13:             **end if**
14:         **end for**
15:         $t \leftarrow empty\_tree()$
16:         **for** $v \in values\_of(e[bv])$ **do**   ▷ Split set by the selected variable
17:             $sse \leftarrow subset(e, bv = v)$
18:             $branch(t, "bv = v") \leftarrow Least.Splits(sse)$ ▷ Redo for each split
19:         **end for**
20:         **return** t ▷ Return sub_tree for selected variable
21:     **else**
22:         **return** $prediction(e)$ ▷ Return prediction value at leaf level
23:     **end if**
24: **end function**

changes on execution time, recursively. E.g. after determining which variable separates better the slow configurations from the faster ones, the algorithm fixes this variable and repeats for each of its distinct values.

As an example, depicted in Fig. 4, we select an on-premise cluster formed by 3 data-nodes with 12 cores per VM and 128 GB RAM, and we want to observe the relevance of variables *disk* (local SSD and HDD), *network* (Infiniband and Ethernet), *IO file buffer* (64 KB and 128 KB) and *block size* (128, 256) for the benchmark *terasort*, fixing then the other variables (*maps* = 4, *sort factor* = 10, no compression and 1 replica). We train a model for this benchmark and predict all the configurations available for the given scenario. Then, using the dichotomous *Least Splits* algorithm we get the tree of relevant variables.

Net	Disk	IO.FBuf	Blk.Size	Prediction (s)
ETH	HDD	65536	128	2249.766
IB	HDD	65536	128	2737.112
ETH	SSD	65536	128	1036.366
IB	SSD	65536	128	1036.366
ETH	HDD	131072	128	2165.927
IB	HDD	131072	128	2653.273
ETH	SSD	131072	128	969.537
IB	SSD	131072	128	969.537
ETH	HDD	65536	256	2249.766
IB	HDD	65536	256	2737.112
ETH	SSD	65536	256	1036.366
IB	SSD	65536	256	1036.366
ETH	HDD	131072	256	2165.927
IB	HDD	131072	256	2653.273
ETH	SSD	131072	256	969.537
IB	SSD	131072	256	969.537
Terasort, 4 maps, sort factor 10, no comp				

```
Disk=SSD
    IO.FBuf=131072 -> 970s
    IO.FBuf=65536 -> 1036s
Disk=HDD
    Net=ETH
        IO.FBuf=131072 -> 2166s
        IO.FBuf=65536 -> 2250s
    Net=IB
        IO.FBuf=131072
            Blk.size=128 -> 2653s
            Blk.size=256 -> 2653s
        IO.FBuf=65536
            Blk.size=128 -> 2737s
            Blk.size=256 -> 2737s
```

Fig. 4. Example of estimation of the selected space of search, with the corresponding descriptive tree

The example shown in Fig. 4 is just one of all the explorations that the ALOJA-ML tool can realize using learned models. E.g., in this example we observe that the variable that defines a division between slower and faster executions is the type of storage units, and then for SSDs only the File Buffer seems relevant, while for HDDs the type of network is the second important variable, then File Buffers and Block Sizes.

As just said, this is an example of what we can experiment and obtain with automatic benchmark modeling using machine learning. This tool and its datasets are open to the community at http://hadoop.bsc.es, also the complete framework can be downloaded and deployed locally at https://github.com/Aloja/aloja for everyone to work with their own data.

5   Conclusions

This article delineated the evolution of ALOJA's focus approach over the last two years to automatically characterize Hadoop deployments and gather performance insights that can improve the execution of current clusters, as well as we aim to influence the design of new cost-effective, data processing infrastructures. On our path from low-level profiling to Predictive Analytics (PA)—our current frontier, we have found specific use cases for the tested performance extraction techniques; as well as finding the data sizes of the different techniques.

Our results show that detailed performance counter collection accounts for over 99% of the data produced, while summaries from the already give the most value to rapidly obtaining cost and performance insights from the data. Aggregated summaries and also allow us to use PA techniques with more ease, and the first results looks promising to speed up and reduce costs of the benchmarking efforts as well to automate. In this case of study we showed how we can model easily separate benchmarks, but the aim is to achieve in the next future a general model that include all benchmarks by studying ways to parametrize those and being able to compare (and then discriminate) them. Also we showed, from our machine learning tool in ALOJA project, a quick algorithm to display how variables affect Hadoop benchmarks execution time, based in the learned models.

Our project is now focused on study how to treat benchmarks and clusters in general models, finding methods to characterize them towards the challenge of learning general models with more proper accuracy. At the same time supporting ALOJA's goal of automating Knowledge Discovery (KD) and recommendations to users.

We also will like to invite fellow researchers and Big Data practitioners to use on our open source tools to expand on the available analytic tools and public benchmark repository.

Acknowledgements. This work is partially supported the BSC-Microsoft Research Centre, the Spanish Ministry of Education (TIN2012-34557), the MINECO Severo Ochoa Research program (SEV-2011-0067) and the Generalitat de Catalunya (2014-SGR-1051).

References

1. Borthakur, D.: System, the Hadoop distributed file: architecture and design. The Apache Software Foundation (2007). http://hadoop.apache.org/docs/r0.18.0/hdfs_design.pdf
2. BSC. Aloja home page (2015). http://aloja.bsc.es/
3. BSC. Performance tools research group page (2015). http://www.bsc.es/computer-sciences/performance-tools
4. BSC. Administrator privileges on headnode of hdinsight-cluster, May 2015. http://www.postseek.com/meta/bd1cddf3af9c7ce35d147e842a686410
5. Gartner. Predictive analytics, May 2015. http://www.gartner.com/it-glossary/predictive-analytics

6. Guitart, J., Torres, J., Ayguad, E., Oliver, J., Labarta, J.: Java instrumentation suite: accurate analysis of java threaded applications. In: Proceedings of the Second Annual Workshop on Java for HPC, ICS 2000, pp. 15–25 (2000)

7. Heger, D.: Hadoop performance tuning - a pragmatic & iterative approach. DH Technologies (2013)

8. Huang, S., Huang, J., Dai, J., Xie, T., Huang, B.: The HiBench benchmark suite: characterization of the MapReduce-based data analysis. In: 22nd International Conference on Data Engineering Workshops, pp. 41–51 (2010)

9. Kambatla, K., Pathak, A., Pucha, H.: Towards optimizing hadoop provisioning in the cloud. In: Proceedings of the 2009 Conference on Hot Topics in Cloud Computing, HotCloud 2009, Berkeley, CA, USA. USENIX Association (2009)

10. Person, L.: Global hadoop market. Allied market research, March 2014

11. Poggi, N., Carrera, D., Call, A., Mendoza, S., Becerra, Y., Torres, J., Ayguadé, E., Gagliardi, F., Labarta, J., Reinauer, R., Vujic, N., Green, D., Blakeley, J.: ALOJA: a systematic study of hadoop deployment variables to enable automated characterization of cost-effectiveness. In: 2014 IEEE International Conference on Big Data, Big Data 2014, Washington, DC, USA, 27–30 October 2014, pp. 905–913 (2014)

12. Schwartz, B., Zaitsev, P., Tkachenko, V.: High Performance MySQL. O'Reilly Media, Sebastopol (2012)

13. Wikipedia. Predictive analytics, May 2015. http://en.wikipedia.org/wiki/predictive_analytics

14. Zhang, Z., Cherkasova, L., Loo, B.T.: Optimizing cost and performance trade-offs for mapreduce job processing in the cloud. In: 2014 IEEE on Network Operations and Management Symposium (NOMS), pp. 1–8. IEEE (2014)

15. Apache Foundation. Apache Hadoop. http://hadoop.apache.org. Accessed Apr. 2015

16. Berral, J.Ll.: Improved management of data-center systems using machine learning. Ph.D. thesis on Computer Science, November 2013

17. Heger, D.: Hadoop performance tuning. https://hadoop-toolkit.googlecode.com/files/Whitepaper-HadoopPerformanceTuning.pdf. Accessed Jan. 2015

18. Intel Corporation. Intel HiBench, Hadoop benchmark suite. https://github.com/intel-hadoop/HiBench. Accessed Apr. 2015

19. Quinlan, R.J.: Learning with continuous classes. In: 5th Australian Joint Conference on Artificial Intelligence, Singapore, pp. 343–348 (1992)

20. Wang, Y., Witten, I.H.: Induction of model trees for predicting continuous classes. In: Poster Papers of the 9th European Conference on Machine Learning (1997)

Experimental Results

Benchmarking the Availability
and Fault Tolerance of Cassandra

Marten Rosselli[1,2]([✉]), Raik Niemann[1,3], Todor Ivanov[1], Karsten Tolle[1],
and Roberto V. Zicari[1]

[1] Frankfurt Big Data Lab, Goethe University Frankfurt am Main,
Frankfurt, Germany
{rosselli,todor,tolle,zicari}@dbis.cs.uni-frankfurt.de
[2] Accenture Germany, Frankfurt, Germany
marten.rosselli@accenture.com
[3] Institute of Information Systems, University of Applied Science Hof, Hof, Germany
raik.niemann@iisys.de

Abstract. To be able to handle big data workloads, modern *NoSQL* database management systems like *Cassandra* are designed to scale well over multiple machines. However, with each additional machine in a cluster, the likelihood for hardware failure increases. In order to still achieve high availability and fault tolerance, the data needs to be replicated within the cluster. Predictable and stable response times are required by many applications even in the case of a node failure. While Cassandra guarantees high availability, the influence of a node failure on the system performance is still unclear.

In this paper, we therefore focus on the availability and fault tolerance of *Cassandra*. We analyze the impact of a node outage within a *Cassandra* cluster on the throughput and latency for different workloads. Our results show that *Cassandra* is well suited to achieve high availability while preserving table response times in case of a node failure. Especially for read intensive applications that require high availability, *Cassandra* is a good choice.

1 Introduction

The exponential growth of data during the last decade led to the development of new data management systems that scale out well over multiple machines. Although the high number of interconnected machines (nodes) brings benefits such an increased overall storage capacity and load balancing to improve the response times, the likelihood for hardware failure increases with each additional node.

There are multiple publications [7,11,14] that deal with the performance and scalability of clustered data managements systems.

In *Fan et al.* [7] the causes of consistency and the staleness of values returned by read operations applied to Cassandra are studied using the same benchmark (YCSB [3]) that we used. *Kuhlenkamp et al.* [11] also show a nearly linear scalability of HBase [8] and Cassandra. Their evaluation of the elasticity of HBase

© Springer International Publishing AG 2016
T. Rabl et al. (Eds.): WBDB 2015, LNCS 10044, pp. 87–95, 2016.
DOI: 10.1007/978-3-319-49748-8_5

and Cassandra showed a tradeoff between two conflicting elasticity objectives, namely the speed of scaling and the performance variability while scaling. *Rabl et al.* [14] focused on the performance and scalability of various key-value stores (e.g. Cassandra, HBase, Voldemort). They observed a linear scalability for Cassandra in most of their tests. Cassandras throughput dominated in all the tests but its latency was high as well compared to the other key-value stores used in this study. In a study by *Beyer et al.* [1] the availability of Cassandra was tested using only a two-node cluster and a simple test case, in which they verified that the database is still available after a node failure.

In this paper, we focus on the availability and fault tolerance of the data management system *Cassandra*. We add to the existing work by focusing on the effects of a node failure within the cluster. We use a bigger seven node cluster and monitored the system throughput and latency in detail throughout our experiments. We performed experiments using the *Yahoo! Cloud Serving Benchmark* (YCSB) on a *Cassandra* cluster. During the workload execution, an outage of a cluster node was initiated while observing the system behavior.

The paper is organized as follows: the remaining part of this section briefly describes the data management system *Cassandra* and the YCSB. Section 2 depicts the test methodology and the hard- and software setup. The experimental results are presented in Sect. 3. The last section summarizes our findings and describes future work in the direction of the availability of data management systems.

1.1   Cassandra Database Management System

Cassandra[1] [12] is a wide-column NoSQL store [2] designed to scale out well across thousands of nodes to handle big data workloads.

Because of its peer-to-peer ring architecture, inspired by *Amazon Dynamo* [6], there is no single point of failure within a *Cassandra* cluster. The data can be replicated multiple times for high availability and fault tolerance.

Cassandra offers a tuneable consistency model where the administrator can choose between availability and fault tolerance on the one hand and data consistency on the other hand. The consistency level is determined by the number of nodes who have to commit a write request [4].

For this study we used the *DataStax Enterprise* (DSE) distribution [5]. This distribution is built on top of *Apache Cassandra* and consists of the *DataStax Enterprise Server* and *DataStax OpsCenter*. While the *Enterprise Server* is based on *Apache Cassandra* and provides additional functionalities such as data security and search capabilities, the *OpsCenter* is a tool for the administration and monitoring of a *Cassandra* cluster.

1.2   Yahoo! Cloud Serving Benchmark

The *Yahoo! Cloud Serving Benchmark* (YCSB) framework [3] was designed for the evaluation of cloud data serving systems using OLTP workloads. Unlike big

---

[1] See http://cassandra.apache.org.

data OLAP benchmarks like *HiBench* [10], *TPCx-HS* [13] or BigBench [9], YCSB was specifically designed to stress test operational cloud data serving systems used for transaction handling. The main components of the YCSB client are the workload generator and a generic database interface. The generic interface can be extended by a programmer to create new database connectors for relational or *NoSQL* databases which have their own query language. A YCSB workload consists of either a single or a mix of the following operation types:

- **Insert**: Insertion of a new record
- **Update**: Replacement of a single field value of an existing record
- **Read**: Reading of either one random record field or of all fields (configurable)
- **Scan**: Scanning of a range of records

In case of a mixed workload, the processing order of the operations is randomly determined by the workload generator. The decisions which record to request are governed by the specified distribution (e.g. uniform, zipfian). By using multiple YCSB client threads, the operations of a workload can be run in parallel.

2   Methodology and Setup

2.1   Workload Definition and Configuration

Using the YCSB data generator, 200 million data records with a record size of one kilobyte were generated. The *Cassandra* replication factor was set to three, resulting in 600 million records stored in the cluster. A single record consisted of ten fields plus a field for the primary key.

The data was uniformly distributed among the cluster nodes. The replication strategy parameter of *Cassandra* was set to `SimpleStrategy`. This strategy places the first replica of an insert operation on node determined by *Cassandra*'s partitioner component. Additional replicas are stored clockwise among the remaining nodes without taking the node's location into account [4].

The write consistency parameter of *Cassandra* was set to "one". This means that a write operation must be successfully written to the commit log and memtable of at least one replica node. The data will eventually become consistent. The read consistency parameter was also set to "one" and the data is returned from the closest replica. These settings provide the highest availability of the database at the cost of a strong consistency [4].

To analyze the behavior of a fully saturated cluster after a node outage, the number of YCSB client threads was set to 1500. At this point the cluster was fully saturated in terms of throughput (operations per second) for all workloads. By decreasing the number of threads the latency for a single thread can be improved for the cost of the throughput.

To be able trace back the behavior of *Cassandra* to specific operation types we used a read and an update heavy workload. Furthermore, to simulate a typical *Cassandra* use case, a workload composed of a mix of both operation types was used. This mixed workload simulates a session store recording recent actions in a web application user session as was defined by *Cooper et al.* [3]. The three workloads are summarized as below:

- **Read** workload: this workload consists of ten million read operations. The benchmark was configured to always read all fields of a record. This workload is similar to the one used by *Cooper et al.* [3], but we are using a uniform instead of a zipfian distribution of the data within the cluster. Cassandra being used as fast user profile cache, where profiles are constructed e.g. in Hadoop, would be a typical use case.
- **Update** workload: this workload consists of ten million update operations. Each update operation replaces a single field value of an existing record. This workload was added in order to evaluate the system performance of Cassandra in contrast to the *read*-only workload.
- **Mixed** workload: this workload consists of five million read and five million update operations. It was designed to simulate a session store recording recent user actions as a typical *Cassandra* use case.

All experiments were repeated three times and the average values were calculated as the representative. After a so called "ramp up time" of the *Cassandra* cluster, one node was shut down to simulate a failure during the workload execution phase. Before each test run the page cache was dropped and the database was cleared from previous data sets. Each test run consists of three phases:

- "Ramp up phase" (100 s): The ramp up phase allows the cluster to reach a stable performance before the performance evaluated further.
- "Pre-failure phase" (80 s): In this phase the system performance was measured before the occurrence of a node failure.
- "Post-failure phase" (120 s): In this phase the system performance was measured after the occurrence of a node failure.

2.2   Hardware and Software Setup

A *Fujitsu BX620S3* blade center was used for all experiments. Table 1 depicts the hard- and software characteristics of the machines used. Seven uniform server blades, further called *Cassandra* nodes, were used to form a *Cassandra* cluster. The cluster was set up using the *OpsCenter* tool. Using its installation routine, all required software packages were automatically installed on the nodes. All unnecessary services of the operating system were turned off.

A separate blade server was used for the benchmark client to generate the load. To ensure this blade server is not the bottleneck, we chose a more powerful hardware configuration. We observed that the machine was never fully saturated and could handle the load.

Since the *Cassandra* connector included in the YCSB package is not compatible with *Cassandra* 2.x, we used an updated binding[2] based on the *Cassandra Query Language* (CQL), the standard way to access *Cassandra* since version 1.2.

---

[2] YCSB *Cassandra* binding based on CQL: https://github.com/jbellis/YCSB.

Table 1. Hardware and software characteristics

Component	Cassandra node	YCSB client
CPU	*AMD Opteron 870*	*AMD Opteron 890*
Main memory	16 GByte DDR-2 reg.	32 GByte DDR-R reg.
Hard drives	2 × 146 GByte (RAID-0)	2 × 300 GByte (RAID-0)
NIC	*Broadcom NetXtreme* BCM5704S, 1 GBit/s transfer speed	
Operating system	*Ubuntu server 12.04 64bit*	
Cassandra	*DataStax Enterprise Server* v4.5.1 (*Apache Cassandra* v2.0.8.39)	
YCSB	v0.1.4 with a CQL-based *Cassandra* binding	
JRE	*Oracle Java Runtime Environment v1.7.60*	

3   Experimental Results and Evaluation

The experimental results for the **read** workload are depicted in Fig. 1 for the throughput and in Fig. 2 for the latency, respectively. After the simulated outage of one *Cassandra* node, the throughput decreased by 7.5% on average whereas the latency increased by 8.6% on average.

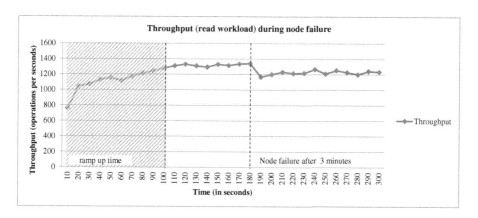

Fig. 1. Mean throughput for the **read** workload during node failure

For the **update** workload, the experimental results regarding the throughput and latency are shown in Figs. 3 and 4. After the outage of one *Cassandra* node, the throughput rate decreased by 10.2% and the latency increased by 11.9%. The effect of the node failure is higher for update operations than for read operations. Since the throughput

The effects of an outage of one *Cassandra* node on the throughput and latency during the **mixed** workload are shown in Figs. 5 and 6. The former figure shows a decrease of the mean throughput by 8.7%. Interestingly the latencies for the read

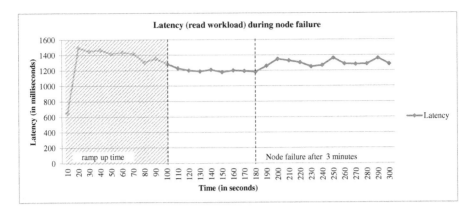

Fig. 2. Mean latency for the `read` workload during node failure

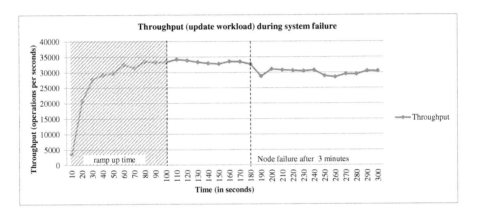

Fig. 3. Mean throughput for the `update` workload during node failure

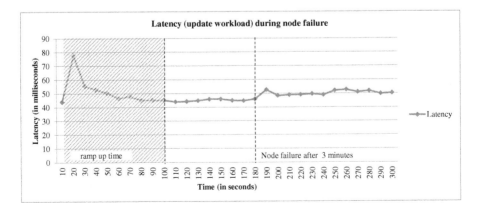

Fig. 4. Mean latency for the `update` workload during node failure

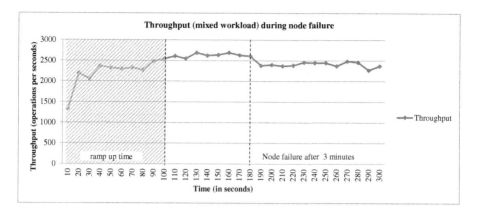

Fig. 5. Mean throughput for the `mixed` workload during node failure

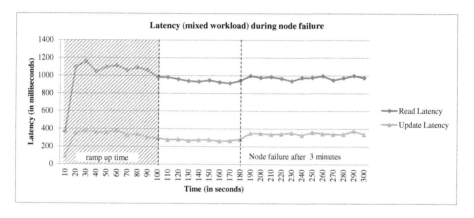

Fig. 6. Mean latency for the `mixed` workload during node failure

and update operations differ significantly in the `mixed` workload. As depicted in Fig. 6, the latency for the read operations increased only by 3.6% on average (5% less compared to the *read* workload) whereas the one for the update operations increased by 27.1% on average (16.2% more compared to the *update* workload).

This can be explained by the different behavior of *Cassandra* for the read and update operation types. As visible in Figs. 1, 2, 3 and 4 *Cassandra* performed much better in terms of throughput and latency for update operations than for read operations. Since there are less read operations concurrently accessing the cluster in the mixed workload (the processing order of the read and update operations is randomly chosen by the workload generator) the average latency for read operations decreases. At the same time the update operations of the mixed workload are concurrently performed with the read operations which negatively affects the update latency. Because the latency is already lower for read operations and higher for update operations before the node outage, the changes afterwards are different compared to the read and *update* workloads.

4   Conclusions and Future Work

It is important to note that *Cassandra* remained stable for all workloads despite a forced node outage. The cluster was still available and the workload execution was not interrupted.

The throughput dropped by 7.5% for the `read` workload whereas the latency increased by 8.6%. For the `update` workload the influence of a node outage was slightly higher: the throughput decreased by 10.2% and the latency increased by 11.9%. The results show that update operations are more negatively affected by a node outage compared to read operations.

For the `mixed` workload the throughput decreased by 8.7%. This is close to the average value based on the results of the read and update workloads. Regarding the latency, the results are different. They show that read operations have a negative impact on concurrent update operations in terms of latency. After a node outage this becomes even more visible: the update latency increased by 27.1% and the read latency increased only by 3.6%. Thus, in a mixed workload, read operations are less affected and the update operations are slowed down by the reads.

For all tested workloads *Cassandra* proved to be well suited to achieve high availability and fault tolerance. Especially for read intensive applications that require high availability all the time, *Cassandra* is a good choice. System administrators could use this study to predict the impact of a node failure on the response times of the end users accordingly to the type of the application workload.

In the future, we plan to benchmark additional *NoSQL* data management systems and extend our results with more workloads and the simulation of multiple nodes failures. The additional systems could either have different data models, for example document oriented (e.g. *MongoDB*) or key-value (e.g. *Riak* with peer-to-peer architecture similar to *Cassandra*).

References

1. Beyer, F., Koschel, A., Schulz, C., Schäfer, M., Astrova, I., Grivas, S.G., Schaaf, M., Reich, A.: Testing the suitability of cassandra for cloud computing environments. In: CLOUD COMPUTING 2011, The Second International Conference on Cloud Computing, GRIDs, and Virtualization, pp. 86–91 (2011)
2. Cattell, R.: Scalable sql and nosql data stores. ACM SIGMOD Rec. **39**(4), 12–27 (2011)
3. Cooper, B.F., Silberstein, A., Tam, E., Ramakrishnan, R., Sears, R.: Benchmarking cloud serving systems with ycsb. In: Proceedings of the 1st ACM Symposium on Cloud Computing, pp. 143–154. ACM (2010)
4. DataStax, Inc.: Datastax cassandra documentation (2015). http://www.datastax.com/docs

5. DataStax, Inc.: Datastax enterprise cassandra distribution (2015). http://www.datastax.com
6. DeCandia, G., Hastorun, D., Jampani, M., Kakulapati, G., Lakshman, A., Pilchin, A., Sivasubramanian, S., Vosshall, P., Vogels, W.: Dynamo: amazon's highly available key-value store. ACM SIGOPS Oper. Syst. Rev. **41**, 205–220 (2007). ACM
7. Fan, H., Ramaraju, A., McKenzie, M., Golab, W., Wong, B.: Understanding the causes of consistency anomalies in apache cassandra. Proc. VLDB Endowment **8**(7), 810–813 (2015)
8. George, L.: HBase: The Definitive Guide. O'Reilly Media Inc., Sebastopol (2011)
9. Ghazal, A., Rabl, T., Hu, M., Raab, F., Poess, M., Crolotte, A., Jacobsen, H.A.: Bigbench: towards an industry standard benchmark for big data analytics. In: Proceedings of the 2013 ACM SIGMOD International Conference on Management of Data, pp. 1197–1208. ACM (2013)
10. Huang, S., Huang, J., Dai, J., Xie, T., Huang, B.: The hibench benchmark suite: characterization of the mapreduce-based data analysis. In: 2010 IEEE 26th International Conference on Data Engineering Workshops (ICDEW), pp. 41–51. IEEE (2010)
11. Kuhlenkamp, J., Klems, M., Röss, O.: Benchmarking scalability and elasticity of distributed database systems. Proc. VLDB Endowment **7**(13), 1219–1230 (2014)
12. Lakshman, A., Malik, P.: Cassandra: a decentralized structured storage system. ACM SIGOPS Oper. Syst. Rev. **44**(2), 35–40 (2010)
13. Nambiar, R., Poess, M., Dey, A., Cao, P., Magdon-Ismail, T., Bond, A., et al.: Introducing tpcx-hs: the first industry standard for benchmarking big data systems. In: Nambiar, R., Poess, M. (eds.) Performance Characterization and Benchmarking. Traditional to Big Data. LNCS, vol. 8904, pp. 1–12. Springer, Switzerland (2014)
14. Rabl, T., Gómez-Villamor, S., Sadoghi, M., Muntés-Mulero, V., Jacobsen, H.A., Mankovskii, S.: Solving big data challenges for enterprise application performance management. Proc. VLDB Endowment **5**(12), 1724–1735 (2012)

Performance Evaluation of Spark SQL Using BigBench

Todor Ivanov[⊠] and Max-Georg Beer

Frankfurt Big Data Lab, Goethe University Frankfurt am Main,
Frankfurt am Main, Germany
{todor,max-georg}@dbis.cs.uni-frankfurt.de

Abstract. In this paper we present the initial results of our work to execute BigBench on Spark. First, we evaluated the scalability behavior of the existing MapReduce implementation of BigBench. Next, we executed the group of 14 pure HiveQL queries on Spark SQL and compared the results with the respective Hive ones. Our experiments show that: (1) for both Hive and Spark SQL, BigBench queries perform with the increase of the data size on average better than the linear scaling behavior and (2) pure HiveQL queries perform faster on Spark SQL than on Hive.

Keywords: Big Data · Benchmarking · BigBench · Hive · Spark SQL

1 Introduction

In the recent years, the variety and complexity of Big Data technologies is steadily growing. Both industry and academia are challenged to understand and apply these technologies in an optimal way. To cope with this problem there is a need of new standardized Big Data benchmarks that cover the entire Big Data lifecycle as outlined by multiple studies [1–3]. The first industry standard Big Data benchmark called TPCx-HS [4] was recently released. It is designed to stress test a Hadoop cluster. While the TPCx-HS is a micro-benchmark (highly I/O and network bound), there is still a need of an end-to-end application-level benchmark [5] that tests the analytical capabilities of a Big Data platform. BigBench [6, 7] has been proposed with the specific intention to fulfill this requirements and is currently available for public review as TPCx-BB [8]. It consists of 30 complex queries. 10 queries were taken from the TPC-DS benchmark [9], whereas the remaining 20 queries were based on the prominent business cases of Big Data analytics identified in the McKinsey report [10]. The BigBench's data model, depicted on Fig. 1, was derived from TPC-DS and extended with unstructured and semi-structured data to fully represent the Big Data Variety characteristic. The data generator is an extension of PDGF [11] that allows to generate all those three data types as well as efficiently scale the data for large scale factors. Chowdhury et al. [12] presented a BigBench implementation for the Hadoop ecosystem [13]. All queries are implemented using Apache Hadoop, Hive, Mahout and the Natural Language Processing Toolkit (NLTK). Table 1 summarizes the number and type of queries of this BigBench implementation.

© Springer International Publishing AG 2016
T. Rabl et al. (Eds.): WBDB 2015, LNCS 10044, pp. 96–116, 2016.
DOI: 10.1007/978-3-319-49748-8_6

Recently, Apache Spark [14] has become a popular alternative to the MapReduce framework, promising faster processing and offering analytical capabilities by Spark SQL [15]. BigBench is a technology agnostic Big Data analytical benchmark, which renders it a good candidate to be implemented in Spark and used as a platform evaluation and comparison tool [7].

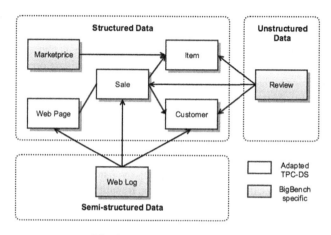

Fig. 1. BigBench schema [12]

Our main objective is to successfully run BigBench on Spark and compare the results with its current MapReduce (MR) implementation [16]. The first step of our work was to execute the largest group of 14 HiveQL queries on Spark. This was possible due to the fact that Spark SQL [15] fully supports the HiveQL syntax.

Table 1. BigBench Queries

Query types	Queries	Number of Queries
Pure HiveQL	Q6, Q7, Q9, Q11, Q12, Q13, Q14, Q15, Q16, Q17, Q21, Q22, Q23, Q24	14
Java MapReduce with HiveQL	Q1, Q2	2
Python Streaming MR with HiveQL	Q3, Q4, Q8, Q29, Q30	5
Mahout (Java MR) with HiveQL	Q5, Q20, Q25, Q26, Q28	5
OpenNLP (Java MR) with HiveQL	Q10, Q18, Q19, Q27	4

In this paper we describe our approach to run BigBench on Spark and present an evaluation of the first experimental results. Our main contributions are:

- Scripts to automate the execution and validation of query results.
- Evaluation of the query scalability on the basis of four different scale factors.
- Comparison of the Hive and Spark SQL query performance.
- Resource utilization analysis of set of seven representative queries.

The remaining of the paper is organized as follows: Sect. 2 describes the necessary steps to implement BigBench on Spark; Sect. 3 discusses the major issues and solutions that we applied during the experiments; Sect. 4 presents the experiments and analyzes the results; Sect. 5 evaluates the queries' resource utilization. Finally, Sect. 6 summarizes the lessons learned and future work.

2 Towards BigBench on Spark

Spark [14] has emerged as a promising general purpose distributed computing framework that extends the MapReduce model by using main memory caching to improve performance. It offers multiple new functionalities such as stream processing (Spark Streaming), machine learning (MLlib), graph processing (GraphX), query processing (Spark SQL) and support for Scala, Java, Python and R programming languages. Due to these new features the BigBench benchmark is a suitable candidate for a Spark implementation, since it consists of queries with very different processing requirements.

Before starting the implementation process, we had to evaluate the different query groups (listed in Table 1) of the available BigBench implementation and identify the adjustments that are necessary. We started by analyzing the largest group of 14 pure HiveQL queries. Fortunately, Spark SQL [15] supports the HiveQL syntax, which allowed us to run this group of queries without any modifications. In Sect. 4, we evaluate the Spark SQL benchmarking results of these queries and compare them with the respective Hive results. It is important to mention that the experimental part of this paper focuses only on Spark SQL and does not evaluate any of the other available Spark components (Spark Streaming, MLlib, GraphX, etc.).

Based on our analysis, we identified multiple steps and open issues that should be completed in order to successfully run all BigBench queries on Spark:

- Re-implementing of the MapReduce jar scripts (in Q1 and Q2) and using external tools like Mahout and OpenNLP running with Spark.
- Making sure that external scripts and files are distributed to Spark executors (Q01, Q02, Q03, Q04, Q08, Q10, Q18, Q19, Q27, Q29, Q30).
- Adjusting the different null value expression from Hive ("\N") to the respective Spark SQL ("null") value (Q3, Q8, Q29, Q30).
- Similar to Hive, new versions of Spark SQL should automatically determine query specific settings, since it is not trivial and very time consuming process.

3   Issues and Improvements

During the query analysis phase, we had to ensure that both the MapReduce and Spark query results are correct and valid. In order to achieve this, the query results should be deterministic and not empty, so that results from query runs of the same scale factor on different platforms are comparable. Furthermore, having an official reference for the data model and result tables (including row counts and sample values) for the various scale factors like the one provided by the TPC-DS benchmark [9] can be helpful for both developers and operators. However, this is not the case with the current MapReduce implementation. The major issue that we encountered were the empty query results, which we solved by adjusting the query parameters, except Q1 (MapReduce) which needs additional changes. By using scripts [16], we collected row counts and sample values for multiple scale factors, which we then used to validate the correctness of the Spark queries. The reference values together with an extended description are provided in our technical report [17].

In spite of our efforts to provide a BigBench reference for result validation, the Mahout queries generate a result text file with varying non-deterministic values. Similarly, the OpenNLP queries generate their results based on randomly generated text attributes, which changed with every new data generation. Validating queries having non-deterministic results is hardly possible.

Finally, we integrated all modifications mentioned above in a setup project that includes a modified version of BigBench 1.0, available on GitHub [16]. In summary, the setup project provides the following benefits: (1) Simplifying commonly used commands like generating and loading data that normally need a lot of unexpressive skip parameters. (2) Running a subset of queries successively. (3) Utilizing the parse-big-bench [18] tool to gather the query execution times in a spreadsheet file even if only a subset of them is executed. (4) Allowing the validation of executed queries by automatically storing row counts and sample row values for every result and Big-Bench's data model table. (5) Improving cleanup of temporary files, i.e. log files created by BigBench.

4   Performance Evaluation

4.1   Experimental Setup

The experiments were performed on a cluster consisting of 4 nodes connected directly through a 1GBit Netgear switch. All 4 nodes are Dell PowerEdge T420 servers. The master node is equipped with 2 × Intel Xeon E5-2420 (1.9 GHz) CPUs each with 6 cores, 32 GB of main memory and 1 TB hard drive. The 3 worker nodes are equipped with 1 × Intel Xeon E5-2420 (2.20 GHz) CPU with 6 cores, 32 GB of RAM and 4 × 1 TB (SATA, 7.2 K RPM, 64 MB Cache) hard drives. Ubuntu Server 14.04.1 LTS was installed on all 4 nodes, allocating the entire first disk. The Cloudera's Hadoop Distribution (CDH) version 5.2.0 was installed on the 4 nodes with the configuration parameters listed in the next section. 8 TB were used in HDFS file system out of the total storage capacity of 13 TB. Due to the small number of cluster nodes, the cluster was configured to work with replication factor of two. The experiments were performed using our modified version of BigBench [16], Hive version 0.13.1 and Spark version 1.4.0-SNAPSHOT (March 27th 2015). A comprehensive description of the experimental environment is available in our report [17].

4.2   Cluster Configuration

Since determining the optimal cluster configuration is very time consuming, our goal was to find a stable one that produces valid query results for the highest tested scale factor (in our case 1000 GB). To achieve this, we applied an iterative approach of executing BigBench queries, adjusting the cluster configuration parameters and validating the query results. First, we started by adapting the *default* CDH configuration to our cluster resources which resulted in a configuration that we called *initial*. After performing a set of tests, we applied the best practices published by Sandy Ryza [19] that were especially relevant for Spark. This resulted in a configuration that we called *final* and was used for the real experiments presented in the next section. Table 2 lists the important parameters for all the three cluster configurations.

Table 2. Cluster Configuration Parameters

Component	Parameter	Default configuration	Initial configuration	Final configuration
YARN	yarn.nodemanager. resource.memory-mb	8 GB	28 GB	*31 GB*
	yarn.scheduler. maximum-allocation-mb	8 GB	28 GB	*31 GB*
	yarn.nodemanager. resource.cpu-vcores	8	8	*11*
Spark	master	local	yarn	*yarn*
	num-executors	2	12	*9*
	executor-cores	1	2	*3*
	executor-memory	1 GB	8 GB	*9 GB*
	spark.serializer	org.apache. spark. serializer. JavaSerializer	org.apache. spark. serializer. JavaSerializer	*org.apache. spark. serializer. KryoSerializer*
MapReduce	mapreduce.map.java. opts.max.heap	788 MB	2 GB	*2 GB*
	mapreduce.reduce.java. opts.max.heap	788 MB	2 GB	*2 GB*
	mapreduce.map. memory.mb	1 GB	3 GB	*3 GB*
	mapreduce.reduce. memory.mb	1 GB	3 GB	*3 GB*
Hive	hive.auto.convert.join (Q9 only)	true	false	*true*
	Client Java Heap Size	256 MB	256 MB	*2 GB*

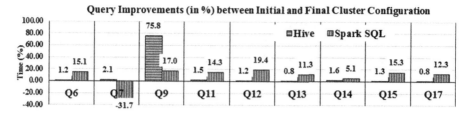

Fig. 2. Improvements between Initial and Final Cluster Configuration for 1 TB data size

Figure 2 depicts the improvements in execution times (in %) between the *initial* and *final* cluster configuration for a set of queries executed on Hive and Spark SQL for 1000 scale factor representing 1 TB data size.

All queries except Q7 benefited from the changes in the *final* configuration. The Hive queries improved on average with 1.3%, except Q9. The reason for Q9 to improve with 76% was that we re-enabled the Hive MapJoins (*hive.auto.convert.join*) and increased the Hive client Java heap size. For the Spark SQL queries, we observed on average an improvement of 13.7%, except Q7 which takes around 32% more time to complete and will be fully investigated in our future work.

4.3 BigBench Data Scalability on MapReduce

In this section we present the experimental results for 4 tested BigBench scale factors (SF): 100 GB, 300 GB, 600 GB and 1000 GB (1 TB). Our cluster used the *final* configuration presented in Table 2. Utilizing our scripts, each BigBench query was executed 3 times and the average value was taken as a representative result, also listed in Table 3. The absolute times for all experiments are available in our technical report [17].

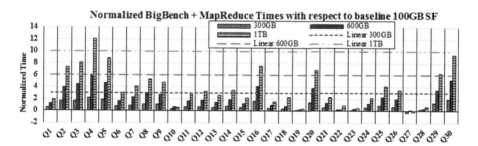

Fig. 3. BigBench + MapReduce Query Times normalized with respect to 100 GB SF

Figure 3 shows all BigBench query execution times for the available MapReduce implementation of BigBench. The presented times for 300 GB, 600 GB and 1 TB are normalized with respect to 100 GB SF as the baseline. Considering the execution times in relation to the different data sizes, we can see that each query differs in its scaling behavior. *Longer normalized times indicate that the execution became slower with the increase of the data size, whereas shorter times indicate better scalability with the increase of the data size.* Q4 has the worst data scaling behavior taking around *2.1 times* longer to process 300 GB, *6 times* longer to process 600 GB and *12 times* longer to process 1 TB data when compared to the 100 GB SF baseline. Q30, Q5 and Q3 show similar scaling behavior. All of them except Q5 (Mahout) are implemented in Python Streaming MR. On the contrary Q27 is almost unchanged (within the range of ±*0.3 times*) with the increase of the data size. Likewise Q19, Q10 implemented in OpenNLP and Q23 in pure HiveQL have slightly worse scaling behaviors.

4.4   BigBench Data Scalability on Spark SQL

This section investigates the scaling behavior of the 14 pure HiveQL BigBench queries executed on Spark SQL using 4 different scale factors (100 GB, 300 GB, 600 GB and 1000 GB). Similar to Fig. 3, the presented times for 300 GB, 600 GB and 1000 GB are normalized with respect to 100 GB scale factor as baseline and depicted on Fig. 4. The average values from the three executions are listed in Table 3.

Table 3. Average query times for the four tested scale factors (100 GB, 300 GB, 600 GB and 1000 GB). The column Δ (%) shows the time difference in % between the baseline 100 GB SFs and the other three SFs for both Hive/MapReduce and Spark SQL.

	Hive/MapReduce							Spark SQL						
SF	100 GB	300 GB		600 GB		1000 GB		100 GB	300 GB		600 GB		1000 GB	
Time	min.	min.	Δ (%)	min.	Δ (%)	min.	Δ (%)	min.	min.	Δ (%)	min.	Δ (%)	min.	Δ (%)
Q1	3.75	5.52	47.2	8.11	116.27	10.48	179.47							
Q2	8.23	21.07	156.01	40.11	387.36	68.12	727.7							
Q3	9.99	26.32	163.46	53.45	435.04	90.55	806.41							
Q4	71.37	221.32	210.1	501.97	603.33	928.68	1201.22							
Q5	27.7	76.56	176.39	155.68	462.02	272.53	883.86							
Q6	6.36	10.69	68.08	16.73	163.05	25.42	299.69	2.54	3.52	38.58	4.83	90.16	6.7	163.78
Q7	9.07	16.92	86.55	29.51	225.36	46.33	410.8	2.54	6.04	137.8	21.47	745.28	41.07	1516.93
Q8	8.59	17.74	106.52	32.46	277.88	53.67	524.8							
Q9	3.13	6.56	109.58	11.5	267.41	17.72	466.13	1.24	1.71	37.9	2.31	86.29	2.82	127.42
Q10	15.44	19.67	27.4	24.29	57.32	22.92	48.45							
Q11	2.88	4.61	60.07	7.46	159.03	11.24	290.28	1.16	1.38	18.97	1.68	44.83	2.07	78.45
Q12	7.04	11.6	64.77	18.67	165.2	29.86	324.15	1.96	3.06	56.12	4.92	151.02	7.56	285.71
Q13	8.38	13	55.13	20.23	141.41	30.18	260.14	2.43	3.59	47.74	5.57	129.22	7.98	228.4
Q14	3.17	5.48	72.87	8.99	183.6	13.84	336.59	1.24	1.56	25.81	2.1	69.35	2.83	128.23
Q15	2.04	3.01	47.55	4.47	119.12	6.37	212.25	1.4	1.59	13.57	1.93	37.86	2.36	68.57
Q16	5.78	14.83	156.57	29.13	403.98	48.85	745.16	3.41	7.88	131.09	23.32	583.87	43.65	1180.06
Q17	7.6	10.91	43.55	14.6	92.11	18.57	144.34	1.56	2.19	40.38	2.91	86.54	3.55	127.56
Q18	8.53	11.02	29.19	14.44	69.28	27.6	223.56							
Q19	6.56	7.22	10.06	7.58	15.55	8.18	24.7							
Q20	8.38	20.29	142.12	39.32	369.21	64.83	673.63							
Q21	4.58	6.89	50.44	10.22	123.14	14.92	225.76	2.68	10.64	297.01	27.18	914.18	48.08	1694.03
Q22	16.64	19.43	16.77	19.82	19.11	29.84	79.33	36.66	60.69	65.55	88.92	142.55	122.68	234.64
Q23	18.2	20.51	12.69	23.22	27.58	25.16	38.24	16.68	27.02	61.99	52.11	212.41	69.01	313.73

(continued)

Table 3. (*continued*)

SF	Hive/MapReduce							Spark SQL						
	100 GB	300 GB		600 GB		1000 GB		100 GB	300 GB		600 GB		1000 GB	
Time	min.	min.	Δ (%)	min.	Δ (%)	min.	Δ (%)	min.	min.	Δ (%)	min.	Δ (%)	min.	Δ (%)
Q24	4.79	7.02	46.56	10.3	115.03	14.75	207.93	3.33	15.27	358.56	42.19	1166.97	77.05	2213.81
Q25	6.23	11.21	79.94	19.99	220.87	31.65	408.03							
Q26	5.19	8.57	65.13	15.08	190.56	22.92	341.62							
Q27	0.91	0.63	-30.77	0.98	7.69	0.7	-23.08							
Q28	18.36	21.24	15.69	24.77	34.91	28.87	57.24							
Q29	5.17	11.73	126.89	22.78	340.62	37.21	619.73							
Q30	19.48	57.68	196.1	119.86	515.3	201.2	932.85							

It is noticeable that Q24 achieves the worst data scalability taking around 3.6 times longer to process 300 GB, 11.7 times longer to process 600 GB and 22 times longer to process 1 TB data when compared to the 100 GB SF baseline. Likewise Q21, Q7 and Q16 have slightly improved data scalability behavior. On the contrary Q15 has the best data scalability taking around 0.14 times for 300 GB, 0.4 times for 600 GB and 0.7 times longer for 1 TB data when compared to the 100 GB SF baseline. Analogously Q11, Q9 and Q14 have slightly worse scalability behavior.

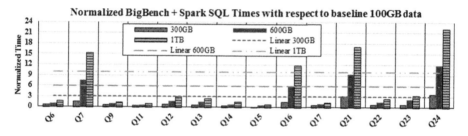

Fig. 4. BigBench + Spark SQL Query Times normalized with respect to 100 GB SF

In summary, our experiments showed that with the increase of the data size the BigBench queries perform on average better than the linear scaling behavior for both the Hive and Spark SQL executions. The only exception for MapReduce is Q4, whereas for Spark SQL these are multiple Q7, Q16, Q21 and Q24. The reason for this behavior probably lies in the reported join issues [20] in the utilized Spark SQL version.

4.5   Hive and Spark SQL Comparison

In addition to the scalability evaluation we compared the query execution time of the 14 pure HiveQL BigBench queries in Hive and Spark SQL with regard to different scale factors.

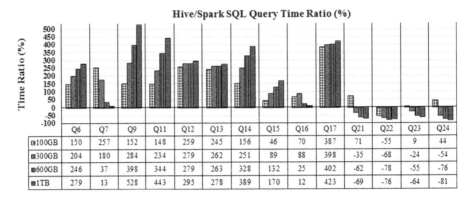

Hive/Spark SQL Query Time Ratio (%)

	Q6	Q7	Q9	Q11	Q12	Q13	Q14	Q15	Q16	Q17	Q21	Q22	Q23	Q24
100GB	150	257	152	148	259	245	156	46	70	387	71	-55	9	44
300GB	204	180	284	234	279	262	251	89	88	398	-35	-68	-24	-54
600GB	246	37	398	344	279	263	328	132	25	402	-62	-78	-55	-76
1TB	279	13	528	443	295	278	389	170	12	423	-69	-76	-64	-81

Fig. 5. Hive to Spark SQL Query Time Ratio defined as *((HiveTime*100)/SparkTime) − 100)*

Figure 5 shows the Hive to Spark SQL query time ratio in % defined as *((HiveTime * 100)/SparkTime) − 100)*. *Positive values indicate faster Spark SQL query execution compared to the Hive ones, whereas negative values indicate slower Spark SQL execution in comparison to Hive.* This figure illustrates that for Q6, Q9, Q11, Q14 and Q15 Spark SQL performs between 46% and 528% faster than Hive.

It is noticeable that this difference increases with a higher data size. For Q12, Q13 and Q17, we observed that the Spark SQL execution times raise slower with the increase of the data sizes, compared to the previous group of queries. On the contrary Q7, Q16, Q21, Q22, Q23 and Q24 drastically increase their Spark SQL execution time for the larger data sets. This results in a declining query time ratio. A highly probable reason for this behavior can be the reported join issue [20] in the utilized Spark SQL version.

5   Query Resource Utilization

This section analyses the resource utilization of a set of representative queries, which are selected based on their behavior presented in the previous section. The first part evaluates the resource utilization of four queries (Q4, Q5, Q18 and Q27) executed on MapReduce, whereas the second compares three HiveQL queries (Q7, Q9 and Q24) executed on both Hive and Spark SQL. The presented metrics (CPU utilization, disk I/O, memory utilization and network I/O) are gathered using the Intel's Performance Analysis Tool (PAT) [21] while executing the queries with 1 TB data size. A full summary of the measured results is available in our technical report [17].

5.1   MapReduce Queries

Queries Q4, Q5, Q18 and Q27 are selected for further analysis based on their scalability behavior and implementation details (Mahout, Python Streaming and OpenNLP).

BigBench's Q4 is chosen for resource evaluation because it is both the slowest of all 30 queries and also shows the worst data scaling behavior on MapReduce. It performs a shopping cart abandonment analysis: For users who added products in their shopping carts but did not check out in the online store, find the average number of pages they visited during their sessions [22]. The query is implemented in HiveQL and executes additional python scripts.

Analogously Q5 was chosen because it is implemented in both HiveQL and Mahout. It builds a model using logistic regression: Based on existing users online activities and demographics, for a visitor to an online store, predict the visitors likelihood to be interested in a given category [22].

Next, we selected Q27 as it showed an almost unchanged behavior when executed with different data sizes. It extracts competitor product and model names (if any) from online product reviews for a given product [22]. The query is implemented in HiveQL and uses the Apache OpenNLP machine learning library for natural language text processing [23]. In order to ensure that this behavior is not caused by the use of the OpenNLP library, Q18 using text processing was selected for resource evaluation. It identifies the stores with flat or declining sales in three consecutive months and check if there are any negative reviews regarding these stores available online [22].

The average values of the measured metrics are shown in Table 4 for all four MapReduce queries. Additionally, the detailed figures for the CPU (Fig. 6), network (Fig. 7) and disk utilization (Fig. 8) in relation to the execution time for the four evaluated queries are included in the Appendix.

It can be observed that Q4 has the highest memory utilization (around 96%) and the highest I/O wait time (around 5%), meaning that the CPU is blocked to wait for the result of outstanding disk I/O requests. The query also has the highest number of context switches per second on average as well as the highest I/O latency time. Both factors are an indication for memory swapping causing massive I/O operations. Taking into account all of the above described metrics, it is no surprise that Q4 is the slowest of all the 30 BigBench queries.

Regarding Q5, it has the highest network traffic (around 8–9 MB/sec) and the highest number of read request per second compared to the other three queries. It is also utilizing around 92% of the memory. Interestingly, the Mahout execution starts after 259 min (15 536 s) in the Q5 execution. It takes only around 18 min and utilizes very few resources in comparison to the HiveQL part of the query. Similar to Q5, Q18 is also memory bound with around 90% utilization. However, it has the highest CPU usage (around 56%) and the lowest I/O wait time (only around 0.30%) compared to the other three queries.

Finally, Q27 shows that the system remains underutilized with only 10% CPU and 27% memory usage during the entire query execution. Further investigation into the query showed that it operated on a very small data set, which slightly varies with the increase of the scale factor. This fact together with the short execution time (just under a minute), render Q27 inappropriate for testing the data scalability and resource utilization of a Big Data platform. It can be used in cases where functional tests involving the OpenNLP library are required.

Table 4. Average Resource Utilization of queries Q4, Q5, Q18 and Q27 on Hive/MapReduce for scale factor 1 TB.

Query		Q4 (Python Streaming)	Q5 (Mahout)	Q18 (OpenNLP)	Q27 (OpenNLP)
Average Runtime (minutes):		928.68	272.53	27.6	0.7
Avg. CPU Utilization %	User	48.82%	51.50%	55.99%	10.03%
	System	3.31%	3.37%	2.04%	1.94%
	I/O wait	4.98%	3.65%	0.3%	1.29%
Memory Utilization %		95.99%	91.85%	90.22%	27.19%
Avg. Kbytes Transmitted per Second		7128.3	8329.02	2302.81	1547.15
Avg. Kbytes Received per Second		7129.75	8332.22	2303.59	1547.14
Avg. Context Switches per Second		11364.64	9859	6751.68	5952.83
Avg. Kbytes Read per Second		3487.38	3438.94	1592.41	1692.01
Avg. Kbytes Written per Second		5607.87	5568.18	988.08	181.19
Avg. Read Requests per Second		47.81	67.41	4.86	14.25
Avg. Write Requests per Second		12.88	13.12	4.66	2.36
Avg. I/O Latencies in Milliseconds		115.24	82.12	20.68	8.89

5.2   Hive and Spark SQL Query Comparison

In this part three HiveQL queries (Q7, Q9 and Q24) are evaluated with the goal to compare the resource utilization of Hive and Spark SQL.

First, we chose BigBench's Q24 because it showed the worst data scaling behavior on Spark SQL. The query measures the effect of competitors' prices on products' in-store and online sales for a given product [22] (Compute the cross-price elasticity of demand for a given product).

Next, Q7 was selected as it sharply decreased its Hive to Spark SQL ratio with the increase of the data size, as depicted on Fig. 5. BigBench's Q7 lists all the stores with at least 10 customers who bought products with the price tag at least 20% higher than the average price of products in the same category during a given month [22]. It was adopted from query 6 of the TPC-DS benchmark [9].

Finally, Q9 was chosen as it showed the highest Hive to Spark SQL ratio difference with the increase of the data size. BigBench's Q9 calculates the total sales for different types of customers (e.g. based on marital status, education status), sales price and different combinations of state and sales profit [22]. It was adopted from query 48 of the TPC-DS benchmark [9].

The average values of the measured metrics are shown in Table 5 for both Hive and Spark SQL together with a comparison represented in the Ratio (%) column. In addition to this, the figures in the appendix depict the resource utilization metrics (CPU utilization, network I/O, disk bandwidth and I/O latencies) in relation to the query's runtime for Q7 (Fig. 9), Q9 (Fig. 10) and Q24 (Fig. 11) for both Hive and Spark SQL with 1 TB data size.

Analyzing the metrics gathered for Q7, it is observable that the Spark SQL execution is only 13% faster than the Hive for 1 TB data size, although for 100 GB this difference was around 256%. This can be explained with the 3 times lower CPU utilization and the higher I/O wait time (around 21%) of Spark SQL. Also the average network I/O (around 3.4 MB/s) of Spark SQL is much smaller than the one of Hive (11.6 MB/s). Interestingly, the standard deviation of the three runs was around 14% for the 600 GB data set and around 4% for the 1 TB data set, which is an indication that the query behavior is not stable. Overall, the poor scaling and unstable behavior of Q7 can be explained with the join issue [20] in the utilized Spark SQL version.

On the contrary, Q9 on Spark SQL is 6.3 times faster than Hive. However, Hive utilizes around 2 times more CPU time and has on average 2.7 times more context switches per second compared with Spark SQL. Both have very similar average network utilization (around 7.5–7.67 MB/s).

Another interesting observation in both queries is that on one hand the average write throughput in Spark SQL is much smaller than its average read throughput. On the other hand, the average write throughput in Hive is much higher than its average read throughput. The reason for this is in the different internal architectures of the engines and in the way they perform I/O operations. It is also important to note that for both queries the average read throughput of Spark SQL is at least 2 times faster than the one of Hive. On the contrary, the average write throughput of Hive is at least 2 times faster than the one of Spark SQL. The reason for this inverse rate lies in the total data sizes that are written and read by both engines.

Finally Q24 executed on Spark SQL is around 5.2 times slower than Hive and represents the HiveQL group of queries with unstable scaling behavior. On Hive, it utilizes on average 49% of the CPU, whereas on Spark SQL the CPU usage is on average 18%. However, for Spark SQL around 11% of the time is spent on waiting for outstanding disk I/O requests (I/O wait), which is much greater than the average for both Hive and Spark SQL. The Spark SQL memory utilization is around 2 times higher than the one of Hive. Similarly, the average number of context switches and the average I/O latency times of Hive are around 20%–23% lower than that of the Spark SQL execution. In this case even the average write throughput of Spark SQL is much higher than the one of Hive. Analogous to Q7, the standard deviation of the three runs was around 8.6% for the 600 GB data set and around 5% for the 1 TB data set, which is a clear sign that the query behavior is not stable. Again the reason is the mentioned join issue [20] in the utilized Spark SQL version.

Table 5. Average Resource Utilization of queries Q7, Q9 and Q24 on Hive and Spark SQL for scale factor 1 TB. The *Ratio* column is defined as *HiveTime/SparkTime* or *SparkTime/HiveTime* and represents the difference between Hive (MapReduce) and Spark SQL for each metric.

Measured metrics		Q7 (HiveQL)			Q9 (HiveQL)		
		Hive	Spark SQL	Hive/Spark SQL Ratio	Hive	Spark SQL	Hive/Spark SQL Ratio
Average Runtime (minutes):		46.33	41.07	**1.13**	17.72	2.82	**6.28**
Avg. CPU Utilization %	User	56.97%	16.65%	**3.42**	60.34%	27.87%	**2.17**
	System	3.89%	2.62%	**1.48**	3.44%	2.22%	**1.55**
	I/O wait	0.40%	21.28%	–	0.38%	4.09%	–
Memory Utilization %		94.33%	93.78%	**1.01**	78.87%	61.27%	**1.29**
Avg. Kbytes Transmitted per Sec.		11650.07	3455.03	**3.37**	7512.13	7690.59	–
Avg. Kbytes Received per Sec.		11654.28	3456.24	**3.37**	7514.87	7691.04	–
Avg. Context Switches per Sec.		10251.24	8693.44	**1.18**	19757.83	7284.11	**2.71**
Avg. Kbytes Read per Sec.		2739.21	6501.03	–	2741.72	13174.12	–
Avg. Kbytes Written per Sec.		7190.15	3364.6	**2.14**	4098.95	1043.45	**3.93**
Avg. Read Requests per Sec.		40.24	66.93	–	9.76	48.91	–
Avg. Write Requests per Sec.		17.13	12.2	**1.4**	10.84	3.62	**2.99**
Avg. I/O Latencies in Millisec.		55.76	32.91	**1.69**	41.67	27.32	**1.53**

Measured metrics		Q24 (HiveQL)		
		Hive	Spark SQL	Spark SQL/Hive Ratio
Average runtime (minutes):		14.75	77.05	**5.22**
Avg. CPU Utilization %	User	48.92%	17.52%	–
	System	2.01%	1.61%	–
	I/O wait	0.48%	11.21%	**23.35**
Memory Utilization %		43.60%	82.84%	**1.9**
Avg. Kbytes Transmitted per Second		3123.24	4373.39	**1.4**
Avg. Kbytes Received per Second		3122.92	4374.41	**1.4**
Avg. Context Switches per Second		7077.1	8821.01	**1.25**
Avg. Kbytes Read per Second		7148.77	7810.38	**1.09**
Avg. Kbytes Written per Second		169.46	3762.42	**22.2**
Avg. Read Requests per Second		22.28	64.38	**2.89**
Avg. Write Requests per Second		4.71	8.29	**1.76**
Avg. I/O Latencies in Milliseconds		21.38	27.66	**1.29**

6   Lessons Learned and Future Work

This paper presented the first results of our initiative to run BigBench on Spark. We started by evaluating the data scalability behavior of the current MapReduce BigBench implementation. The results revealed that a subset of the OpenNLP (MR) queries (Q19, Q10) scale best with the increase of the data size, whereas a subset of the Python Streaming (MR) queries (Q4, Q30, Q3) show the worst scaling behavior. Then we executed the 14 pure HiveQL queries on Spark SQL and compared their execution times with the respective Hive ones. We observed that both Hive and Spark SQL queries achieve on average better than linear data scaling behavior. Our analysis identified a group of unstable queries (Q7, Q16, Q21, Q22, Q23 and Q24), which were influenced by join issue [20] in Spark SQL. For this queries, we observed a much higher standard deviation (4%−20%) between the three executions even for the larger data sizes.

Our experiments showed that for the stable pure HiveQL queries (Q6, Q9, Q11, Q12, Q13, Q14, Q15 and Q17), Spark SQL performs between 1.5 and 6.3 times faster than Hive.

Last but not least, investigating the resource utilization of queries with different scaling behavior showed that the majority of evaluated MapReduce queries (Q4, Q5, Q18, Q7 and Q9) are memory bound. For queries Q7 and Q9, Spark SQL:

- Utilized less CPU, whereas it showed higher I/O wait time than Hive.
- Read more data from disk, whereas it wrote less data than Hive.
- Utilized less memory than Hive.
- Sent less data over the network than Hive.

The next step is to investigate the influence of various data formats (ORC, Parquet, Avro etc.) on the query performance. Another direction to extend the study will be to repeat the experiments on other SQL-on-Hadoop engines.

Acknowledgment. This work has benefited from valuable discussions in the SPEC Research Group's Big Data Working Group. We would like to thank Tilmann Rabl (University of Toronto), John Poelman (IBM), Bhaskar Gowda (Intel), Yi Yao (Intel), Marten Rosselli, Karsten Tolle, Roberto V. Zicari and Raik Niemann of the Frankfurt Big Data Lab for their valuable feedback. We would like to thank the Fields Institute for supporting our visit to the Sixth Workshop on Big Data Benchmarking at the University of Toronto.

A. BigBench Queries' Resource Utilization

See Figs. 6, 7, 8, 9, 10 and 11.

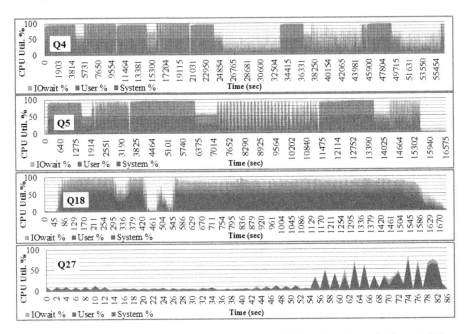

Fig. 6. CPU Utilization of queries Q4, Q5, Q18 and Q27 on Hive for scale factor 1 TB.

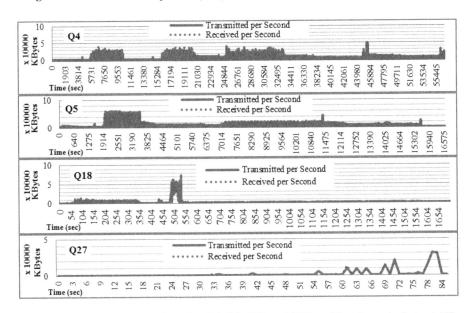

Fig. 7. Network Utilization of queries Q4, Q5, Q18 and Q27 on Hive for scale factor 1 TB.

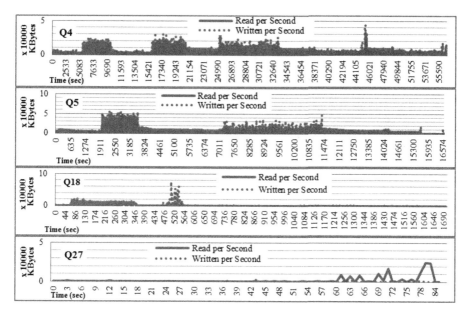

Fig. 8. Disk Utilization of queries Q4, Q5, Q18 and Q27 on Hive for scale factor 1 TB.

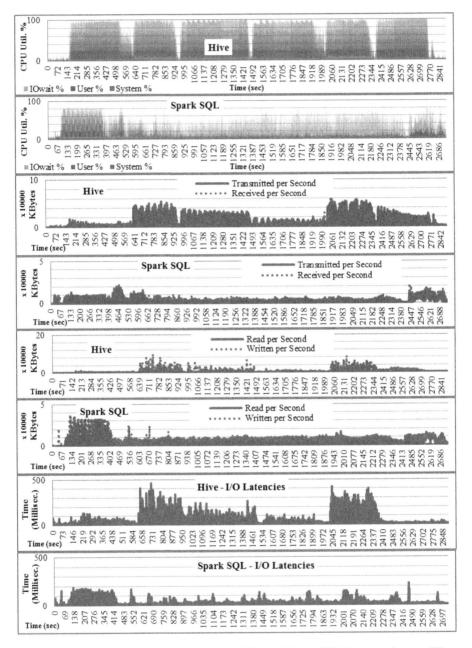

Fig. 9. Resource Utilization of query Q7 on Hive and Spark SQL for scale factor 1 TB.

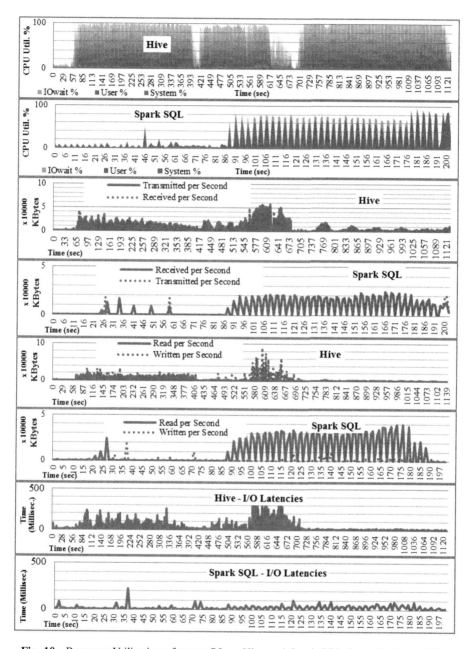

Fig. 10. Resource Utilization of query Q9 on Hive and Spark SQL for scale factor 1 TB.

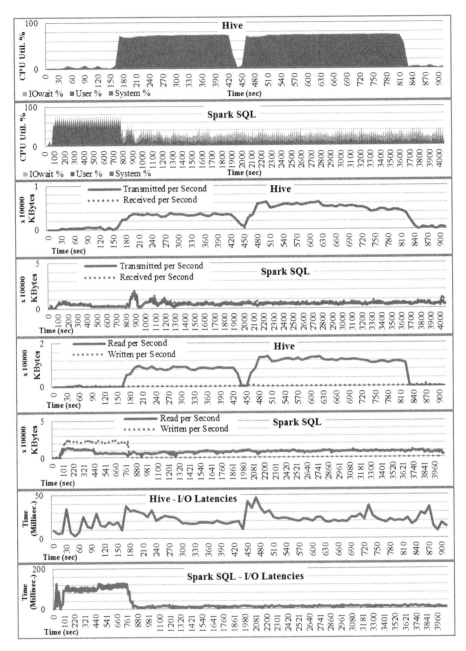

Fig. 11. Resource Utilization of query Q24 on Hive and Spark SQL for scale factor 1 TB.

References

1. Chen, Y.: We don't know enough to make a big data benchmark suite-an academia-industry view. In: Proceeding WBDB, 2012 (2012)
2. Carey, Michael, J.: BDMS performance evaluation: practices, pitfalls, and possibilities. In: Nambiar, R., Poess, M. (eds.) TPCTC 2012. LNCS, vol. 7755, pp. 108–123. Springer, Heidelberg (2013). doi:10.1007/978-3-642-36727-4_8
3. Chen, Y., Raab, F., Katz, R.: From TPC-C to big data benchmarks: a functional workload model. In: Rabl, T., Poess, M., Baru, C., Jacobsen, H.-A. (eds.) WBDB -2012. LNCS, vol. 8163, pp. 28–43. Springer, Heidelberg (2014). doi:10.1007/978-3-642-53974-9_4
4. Nambiar, R., Poess, M., Dey, A., Cao, P., Magdon-Ismail, T., Ren, D.Q., Bond, A.: Introducing TPCx-HS: the first industry standard for benchmarking big data systems. In: Nambiar, R., Poess, M. (eds.) TPCTC 2014. LNCS, vol. 8904, pp. 1–12. Springer, Heidelberg (2014)
5. Baru, C., Bhandarkar, M., Nambiar, R., Poess, M., Rabl, T.: Setting the direction for big data benchmark standards. In: Nambiar, R., Poess, M. (eds.) TPCTC 2012. LNCS, vol. 7755, pp. 197–208. Springer, Heidelberg (2013). doi:10.1007/978-3-642-36727-4_14
6. Ghazal, A., Rabl, T., Hu, M., Raab, F., Poess, M., Crolotte, A., Jacobsen, H.-A.: BigBench: towards an industry standard benchmark for big data analytics. In: Proceedings of the 2013 ACM SIGMOD International Conference on Management of Data, New York, NY, USA, pp. 1197–1208 (2013)
7. Baru, C., et al.: Discussion of BigBench: a proposed industry standard performance benchmark for big data. In: Nambiar, R., Poess, M. (eds.) TPCTC 2014. LNCS, vol. 8904, pp. 44–63. Springer, Heidelberg (2015). doi:10.1007/978-3-319-15350-6_4
8. TPC, "TPCx-BB." http://www.tpc.org/tpcx-bb
9. TPC, "TPC-DS." http://www.tpc.org/tpcds/
10. Manyika, J., Chui, M., Brown, B., Bughin, J., Dobbs, R., Roxburgh, C., Byers, A.H., Big data: the next frontier for innovation, competition, and productivity. McKinsey Glob. Inst., pp. 1–137 (2011)
11. Rabl, T., Frank, M., Sergieh, H.M., Kosch, H.: A data generator for cloud-scale benchmarking. In: Nambiar, R., Poess, M. (eds.) TPCTC 2010. LNCS, vol. 6417, pp. 41–56. Springer, Heidelberg (2011). doi:10.1007/978-3-642-18206-8_4
12. Chowdhury, B., Rabl, T., Saadatpanah, P., Du, J., Jacobsen, H.-A.: A BigBench implementation in the hadoop ecosystem. In: Rabl, T., Jacobsen, H.-A., Raghunath, N., Poess, M., Bhandarkar, M., Baru, C. (eds.) WBDB 2013. LNCS, vol. 8585, pp. 3–18. Springer, Heidelberg (2014). doi:10.1007/978-3-319-10596-3_1
13. Big-Data-Benchmark-for-Big-Bench   GitHub.   https://github.com/intel-hadoop/Big-Data-Benchmark-for-Big-Bench
14. Zaharia, M., Chowdhury, M., Das, T., Dave, A., Ma, J., McCauley, M., Franklin, M., Shenker, S., Stoica, I.: Resilient distributed datasets: a fault-tolerant abstraction for in-memory cluster computing. In: Proceedings of the 9th USENIX conference on Networked Systems Design and Implementation, p. 2 (2012)
15. Armbrust, M., Xin, R.S., Lian, C., Huai, Y., Liu, D., Bradley, J.K., Meng, X., Kaftan, T., Franklin, M.J., Ghodsi, A.: Spark SQL: relational data processing in spark. In: Proceedings of the 2015 ACM SIGMOD International Conference on Management of Data (2015)
16. Frankfurt Big Data Lab, "Big-Bench-Setup GitHub". https://github.com/BigData-Lab-Frankfurt/Big-Bench-Setup
17. Ivanov, T., Beer, M.-G.: Evaluating hive and spark SQL with BigBench, arXiv:1512.08417 (2015)

18. Harsch, T.: Parse-big-bench utility - bitbucket. https://bitbucket.org/tharsch/parse-big-bench
19. Ryza, S.: How-to: tune your apache spark jobs (Part 2) | Cloudera Engineering Blog, 30March 2015
20. Yi Z.: [SPARK-5791] [Spark SQL] show poor performance when multiple table do join operation. https://issues.apache.org/jira/browse/SPARK-5791
21. Intel, "PAT Tool GitHub". https://github.com/intel-hadoop/PAT
22. Rabl, T., Ghazal, A., Hu, M., Crolotte, A., Raab, F., Poess, M., Jacobsen, H.-A.: BigBench specification V0.1. In: Rabl, T., Poess, M., Baru, C., Jacobsen, H.-A. (eds.) WBDB -2012. LNCS, vol. 8163, pp. 164–201. Springer, Heidelberg (2014). doi:10.1007/978-3-642-53974-9_14
23. Apache OpenNLP. https://opennlp.apache.org/

Accelerating BigBench on Hadoop

Yan Tang[1(✉)], Gowda Bhaskar[2(✉)], Jack Chen[1(✉)], Xin Hao[1(✉)],
Yi Zhou[1(✉)], Yi Yao[1(✉)], and Lifeng Wang[1(✉)]

[1] Intel Corporation, Shanghai, China
{yan.a.tang,jack.z.chen,xin.hao,yi.a.zhou,
yi.a.yao,lifeng.a.wang}@intel.com
[2] Intel Corporation, Santa Clara, CA, USA
bhaskar.d.gowda@intel.com

Abstract. Benchmarking Big Data systems is an open challenge. The existing Micro-Benchmarks (e.g. TeraSort) do not present an end-to-end scenario in real world. To solve this issue, a new towards industry standard benchmark for Big Data Analytics called BigBench has been proposed. And with BigBench, we've been keeping our collaboration with Apache Open Source Community to work on performance tuning and optimization for Hadoop ecosystem. In this paper, we share our contributions to BigBench, and present our tuning and optimization experience along with the benchmark results.

Keywords: Big data benchmark · BigBench · TPC · Hadoop · Spark

1 Introduction

Nowadays, enterprises everywhere show a tremendous interest in querying and analyzing the increasing data for enterprise applications support. They need some storage systems and processing paradigms to find the potential significant value in big data. As these systems become mature gradually, there is a common need to evaluate and compare the products' performance. However, the existing benchmarks can't reflect the overall performance for the big data systems. Thus, BigBench [1] was proposed as the first end-to-end benchmark, which has been already approved and published by TPC. When comparing to the Micro Benchmarks, such as HiBench, SparkBench, which are more suitable for regression testing, BigBench is the industry standard for addressing the system shortcomings and providing the price/performance and energy consumption comparison [2].

Hadoop is a popular software framework to process vast amounts of data in parallel on large clusters. Thus, we need to tune our Hadoop based workloads to take full advantage of this framework to get a better performance of your Hadoop cluster. We've been working on performance tuning and optimization for Hadoop ecosystem since 2011. And with BigBench, we can keep collaboration with the Apache Open Source community to improve the performance for big data systems.

In this paper, we present some of our scaling experience with BigBench. In the process of our scaling experiments, we've obtained some main characteristics for each scaling scenario. After some deep investigation, we addressed some performance issues

© Springer International Publishing AG 2016
T. Rabl et al. (Eds.): WBDB 2015, LNCS 10044, pp. 117–127, 2016.
DOI: 10.1007/978-3-319-49748-8_7

in the 30 queries and proposed our optimization. In addition, we provide some of our tuning experience in tuning Hive on Spark [3] with BigBench.

The paper is organized as follows. Section 2 describes some related benchmark and spark tuning efforts in big data systems. Section 3 provides some of our scaling experience with BigBench, including data scaling, cluster nodes scaling and stream scaling. Section 4 introduces our contributions to BigBench query optimization. Section 5 presents some of our tuning experience for Hive on Spark with BigBench. The Spark1.5 Dynamic Allocation Optimization and MapJoin Optimization are described in detail in Sects. 5.1 and 5.2. And Sect. 6 summarizes the paper and suggests our future directions.

2   Related Work

Today, TPC benchmarks are commonly used for big data systems' performance benchmarking. TPC-H [4] and TPC-DS [5] benchmarks, developed by the Transaction Processing Performance Council, are widely used in big data analytics. However, they're both pure SQL benchmarks and can't cover the variety of characteristics of big data systems. Thus, we've moved our focus to BigBench (TPCx-BB), the end-to-end big data benchmark based on TPC-DS.

Several basic tuning guides have been noted in the Spark official website [6], but few papers have been related with Spark performance. Chiba and Onodera presented their JVM parameters tuning for spark performance based on TPC-H queries [7]. And their work is very similar to us, while the workloads and tuning strategy are different.

3   Scaling Experience with BigBench

The focus of our work is on some scaling experiments with BigBench v0.5 in Hive on MapReduce [8] engine, including Data Scaling, Cluster Nodes Scaling and Stream Scaling. We summarized and analyzed the experiments' result to find some trends and characteristics for each scenario.

3.1   Data Scaling

First, we introduce our experiment environment. We conducted the data scaling experiment on an 8-node cluster (including 6 working node). And each node is equipped with 18 cores, 192 GB RAM and 8*3 TB SATA in 10GbE. The system software stack is CentOS6.7 (kernel: 2.6.32-573.el6.x86_64) and Cloudera CDH5.5.1.

Next step is to generate 1T, 3T and 10T dataset on HDFS with BigBench and run three rounds of BigBench to compare all the benchmark results.

Figure 1 shows that while enlarging the data scale from 1T to 10T, the benchmark execution time increased nearly according to the growing proportions of data scale. As the cluster resource is not fully used in small data scale, we can find that the increasing rate is a little smaller than the growing proportions of data scale.

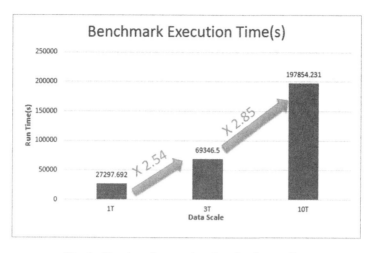

Fig. 1. Benchmark execution time for data scaling

3.2 Cluster Nodes Scaling

First, we introduce our experiment environment. We conducted the cluster nodes scaling experiment based on two clusters, one has 8 nodes (including 6 working node), and the other has 12 nodes (including 10 working node). Each node is equipped with 18 cores, 192 GB RAM and 8*3 TB SATA in 10GbE. The system software stack is CentOS6.7 (kernel: 2.6.32-573.el6.x86_64) and Cloudera CDH5.5.1.

Next, we benchmark the two clusters using 3 TB and 10 TB data scale with BigBench. In theory, when extending the cluster from 6 nodes to 10 nodes, we could get at most 66.7% performance improvement if the cluster is fully utilized. And the higher the cluster resource utilization, the closer improvement to this value. Figure 2

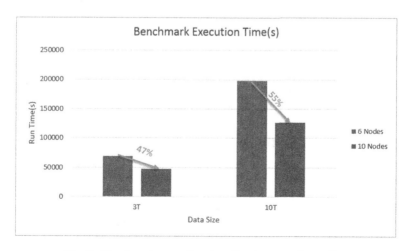

Fig. 2. Benchmark execution time for cluster nodes scaling

shows that in 3T dataset, we can get 47% performance gain after enlarging the cluster nodes, and 55% in 10T dataset, which is closer to the theoretical value.

3.3 Stream Scaling

BigBench has a throughput phase [9], in which all the queries run in parallel streams in different orders. And users can define the stream number by a specific parameter.

First, we introduce our experiment environment. We conducted the stream scaling experiment on a 12-node cluster (including 10 working node). And each node is equipped with 18 cores, 256 GB RAM and 8*3 TB SATA in 10GbE. The system software stack is CentOS6.7 (kernel: 2.6.32-573.el6.x86_64) and Cloudera CDH5.5.1.

Next step is to generate 3 TB dataset and run BigBench with different stream numbers in throughput phase. As seen in Fig. 3, the x axis means the stream number, and the two y axes represent the BigBench execution time and BigBench score, which is generated based on a formula [10]. It shows that while increasing the stream number, the score can keep improving at first and remain the same later when the cluster resource utilization has reached the peak.

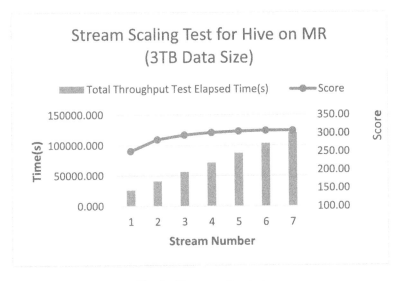

Fig. 3. Stream scaling test

4 BigBench Query Optimization

We have made a deep analysis for all the 30 queries in BigBench, and proposed our optimizations for some of them. Figure 4 shows the performance improvement we've got from the query optimizations using Hive on MapReduce engine in 100 GB, 1 TB and 3 TB data scale. And it can be seen that the performance gain varies between different data size. For example, our optimization produced much more improvement while enlarging the data scale for Q20.

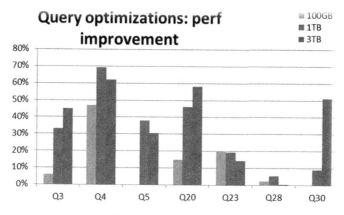

Fig. 4. Query optimizations

Here we pick up Q20 for performance evaluation. Q20 performs customer segmentation for return analysis based on K-means clustering algorithm. It runs hive query to extract K-means input data as the first step, and then leaves it to mahout to do K-means. But we find the Hive query execution time accounts for about 80% of total execution time of the query. So we start to investigate the hive part for optimization opportunities. As shown in Fig. 5, the original Q20 contains two steps. Firstly, it conducts a LEFT OUTER JOIN on store_sales table and store_returns table, then it

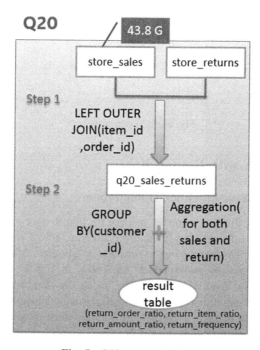

Fig. 5. Q20 processing flow

does GROUPBY and AGGREGATION operation on the temp table to generate the result information, which would be used in machine learning. Unfortunately, the left table store_sales is a very large table which is about 43.8 G in 1 TB testing scale, and it takes a lot of time for the join operation.

To improve this situation, we propose our optimization for Q20. As seen in Fig. 6, the key point is to finish the GROUPBY and AGGREGATION operation before the join operation. Consequently, the execution time of LEFT OUTER JOIN can be reduced a lot since the left table 's2' becomes much smaller. In addition, we use customer id as a key for LEFT OUTER JOIN instead of item id and order id, which makes the query implementation clearer and faster.

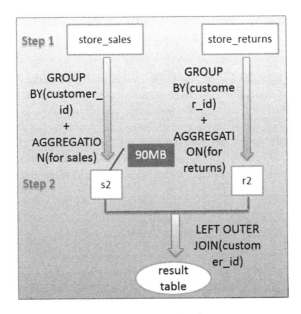

Fig. 6. Q20 optimization

5 Tuning Spark with BigBench

5.1 Spark1.5 Dynamic Allocation Optimization

During our Stream Scaling experiments in Hive on MapReduce engine, we found the cluster resource was fully utilized in multiple streams, where all the streams ran in parallel. However, when moving our experiment to Hive on Spark engine, the performance result was not so as expected. As shown in Fig. 7, the CPU utilization remained in a very low level while increasing the stream number.

After some investigation, we found the spark dynamic allocation [11] feature may solve this issue. Before Spark 1.5, this feature still has some issues. And through some evolution and refinement of codes, now it has become much more mature and robust.

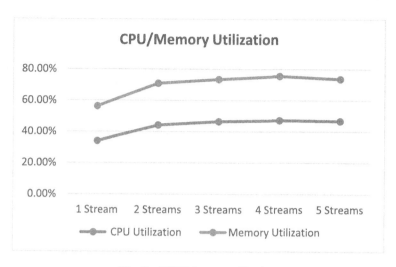

Fig. 7. CPU/Memory utilization

Why It Matters. Spark can adjust the resources for your application based on the workloads after enabling Dynamic Allocation. Comparing to Static Allocation, this mechanism leads to better resource scheduling. Figure 8 shows that the allocated resource is much more than the real used resource in Static Allocation. If your application is the only one workload running on the cluster, it may get better performance due to resource pre-acquisition. However, in reality, it is more common to have multiple users. Thus Static Allocation may introduces some resource scheduling problems.

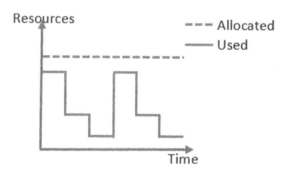

Fig. 8. Static Allocation

The first problem is underutilized of cluster resources. For example, if you start a spark-shell application, but do not submit even one job. However, the resource is still occupied by spark and it will lead to the second bad situation, where the other applications must be queued for resource releasing. The third problem is lack of elastic resource scaling ability. As we know, it is very hard to estimate the needed resources beforehand, even when you're running an iterative workload in which data will be inflated during iteration.

Now the Dynamic Allocation has been introduced to handle these issues. In Dynamic Allocation, spark bridges the executors and the task to calculate the required resources. And it requests and removes the executors at run time. So the number of executors is decided by your workloads. In consequence, the utilization of cluster resources has become more efficient in Dynamic Allocation than that in Static Allocation.

Performance Evaluation. To enable Dynamic Allocation, we need to set the spark property "spark.dynamicAllocation.enabled" to true. Since spark requests and removes the executors at run time, if we still save the status of RDDs to executors, they would be lost. So we must enable the shuffle service to preserve all the shuffle files to nodemanagers by the property "spark.shuffle.service.enabled".

We conducted the tuning experiment on a 12-node cluster (including 10 working node). And each node is equipped with 18 cores, 256 GB RAM and 8*3 TB SATA in 10GbE. The system software stack is CentOS6.7 (kernel: 2.6.32-573.el6.x86_64) and Cloudera CDH5.5.1.

Then we generated 3 TB dataset and ran BigBench with different stream numbers in throughput test. Figure 9 shows that comparing to the former performance result, the CPU utilization can keep improving when increasing stream from 1 to 11 after enabling Dynamic Allocation. And obviously, the resource utilization has also become much more efficient than the original one.

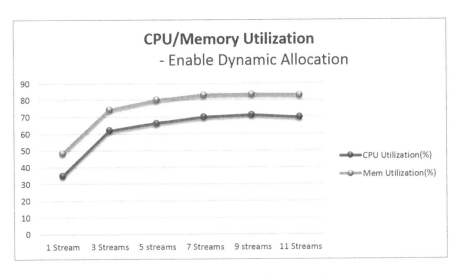

Fig. 9. CPU/Memory utilization of Dynamic Allocation

In addition, as seen in Fig. 10, the BigBench score only stays in a very low level when disabling Dynamic Allocation. And after enabling the feature, the BigBench score can keep improving at first and remains the same later when the stream number is

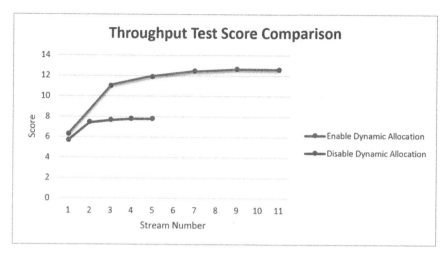

Fig. 10. Throughput test score comparison

enough to full utilize the cluster resources. Obviously, the highest score is much better after enabling the Spark Dynamic Allocation feature.

5.2 MapJoin Optimization for Hive on Spark

Problem Detecting. When joining a fact table with a dimension table, it checks the small table's threshold to tell if the MapJoin is triggered. As we know, MapJoin [12] can broadcast the small table to all the mappers and avoid shuffle phase to improve the performance. For Hive on MapReduce, we've got a reasonable performance result with the default small table threshold which is 25 MB. However, when moving towards Hive on Spark, the performance result was not as expected.

After some investigation, we found the performance issue was resulted from the different Map Join trigger mechanisms between Hive on MapReduce and Hive on Spark. While Hive on MapReduce uses the tables' total size on HDFS to compare with the small table threshold, Hive on Spark uses the tables' raw data size. The raw data size is the size of the original data set, and the total size is the amount of storage it takes.

Table 1 lists the two types of size for all the 23 tables in BigBench with 1 TB data size. As seen in Table 1, the four highlighted items are that enabled Map Join in Hive on MapReduce, but not enabled in Hive on Spark.

Performance Evaluation. To evaluate the performance enhancement, we picked up q05 and q16 to perform the experiments. The two queries have a join operation on the highlighted tables customer_demographics and date_dim. Based on the tables' size list in Table 1, we tune the configuration "hive.mapjoin.smalltable.filesize" to 1 GB.

Table 1. Table size

#	Table (SF=1000)	totalSize	rawDataSize	Map Join enabled? (Hive on MR Default)	Map Join enabled? (Hive on Spark Default)
1	customer	127M	2.6G		
2	customer_address	27M	1.6G		
3	customer_demographics	6.8M	685M	Yes	No
4	date_dim	1.7M	117M	Yes	Yes
5	household_demographics	14K	816K	Yes	Yes
6	income_band	0.4K	0.3K	Yes	Yes
7	inventory	5.6G	99.7G		
8	item	74M	942M		
9	item_marketprices	22M	656M	Yes	No
10	product_reviews	1.2G	3.7G		
11	promotion	217K	7M	Yes	Yes
12	reason	24K	159K	Yes	Yes
13	ship_mode	2K	9K	Yes	Yes
14	store	49K	799K	Yes	Yes
15	store_returns	2.3G	70.6G		
16	store_sales	43.8G	1628G		
17	time_dim	219K	39M	Yes	No
18	warehouse	3K	28.9K	Yes	Yes
19	web_clickstreams	91G	432G		
20	web_page	67.7K	682K	Yes	Yes
21	web_returns	2.7G	72.6G		
22	web_sales	62G	2085G		
23	web_site	6.9K	61K	Yes	Yes

Figure 11 shows q05 can work twice as fast as the default one by enabling Map-Join. And q16 also has about 1.5 times speedup. Obviously, the MapJoin enabling is very helpful for this workload characterization.

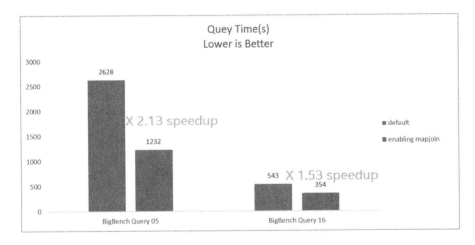

Fig. 11. Q05&Q16 query time

6 Conclusion and Future Work

In this paper we present our experience of BigBench, including scaling experience, query optimization and tuning experience for Spark. Our scaling experience has shown some trends and characteristics for each scaling scenario which is really useful for performance prediction. Our query optimization strategies have outperform up to 70% performance improvement in Q04, and 10%–40% in average in some other queries.

And our dynamic allocation optimization and MapJoin optimization for spark have led to much better performance for Hive on Spark.

In our future work, we will plan to evaluate and tune BigBench on Spark 1.6. Since Spark 1.6, it uses a new memory management, which is implemented as UnifiedMemoryManager, while Spark 1.5 uses Legacy mode [13, 14]. And we'll continue to find the optimization opportunities for Spark 1.6 with BigBench. In addition, for data scaling experiment, we plan to investigate and enable the large data scale, 30 TB and 100 TB in a large cluster.

References

1. TPCx-BB. http://www.tpc.org/tpcx-bb/
2. Ghazal, A., Rabl, T., Hu, M., Raab, F., Poess, M., Crolotte, A., Jacobsen, H.: BigBench: towards an industry standard benchmark for big data analytics. In: SIGMOD (2013)
3. Hive on Spark. https://issues.apache.org/jira/browse/HIVE-7292
4. TPC-H. http://www.tpc.org/tpch/
5. TPC-DS. http://www.tpc.org/tpcds/
6. Tuning Spark. http://spark.apache.org/docs/latest/tuning.html
7. Chiba, T., Onodera, T.: Workload characterization and optimization of TPC-H queries on Apache Spark. Computer Science (2015)
8. Dean, J., Ghemawat, S.: MapReduce: simplified data processing on large clusters. Google Inc (2004)
9. Rabl, T., Frank, M., Danisch, M., Gowda, B., Jacobsen, H.-A.: Towards a complete BigBench implementation. In: Rabl, T., Sachs, K., Poess, M., Baru, C., Jacobson, H.-A. (eds.) WBDB 2015. LNCS, vol. 8991, pp. 3–11. Springer, Heidelberg (2015). doi:10.1007/978-3-319-20233-4_1
10. Baru, C., Bhandarkar, M., Curino, C., Danisch, M., Frank, M., Gowda, B., Jacobsen, H., Jie, H., Kumar, D., Nambiar, R., Poess, M., Raab, F., Rabl, T., Ravi, N., Sachs, K., Sen, S., Yi, L., Youn, C.: Discussion of BigBench: a proposed industry standard performance benchmark for big data
11. Spark Dynamic Allocation. http://spark.apache.org/docs/latest/job-scheduling.html#dynamic-resource-allocation
12. Blanas, S., Patel, M., Ercegovac, V., Rao, J., Shekita, J., Tian, Y.: A comparison of join algorithms for log processing in MapReduce. In: Proceedings of the 2010 ACM SIGMOD International Conference on Management of data, pp. 975–986
13. Spark Memory Management. http://spark.apache.org/docs/latest/configuration.html#memory-management
14. Ganelin, I., Orhian, E., Sasaki, K., York, B.: Spark^{TM}: Big Data Cluster Computing in Production. Wiley, New York (2016). Chap. 2

Author Index

Printed in the United States
By Bookmasters